Progress in Inflammation Research

Series Editor

Prof. Dr. Michael J. Parnham
PLIVA
Research Institute
Prilaz baruna Filipovica 25
10000 Zagreb
Croatia

Forthcoming titles:
Mechanisms and Mediators of Neuropathic Pain, A.B. Malmberg, S.R. Chaplan (Editors), 2002
The Hereditary Basis of Allergic Disorders, J. Holloway, S. Holgate (Editors), 2002
Bone Morphogenetic Proteins: From Laboratory to Clinical Practice, S. Vukicevic, K.T. Sampath (Editors), 2002

(Already published titles see last page.)

Migraine: A Neuroinflammatory Disease?

Egilius L. H. Spierings
Margarita Sánchez del Río

Editors

Springer Basel AG

Editors

Egilius L.H. Spierings
Brigham and Women's Hospital
Harvard Medical School
75 Francis Street
Boston, MA 02115
USA

Margarita Sánchez del Río
Headache Unit
Department of Neurology
Fundación Hospital Alcorcón
Juan Carlos I University
Madrid
Spain

A CIP catalogue record for this book is available from the Library of Congress, Washington D.C., USA

Deutsche Bibliothek Cataloging-in-Publication Data
Migraine: a neuroinflammatory disease? / Egilius L. H. Spierings ; Margarita Sanchez del Rio, ed.. - Basel ; Boston ; Berlin : Birkhäuser, 2002
(Progress in inflammation research)
ISBN 978-3-0348-9449-4 ISBN 978-3-0348-8131-9 (eBook)
DOI 10.1007/978-3-0348-8131-9

Originally published by Birkhäuser Verlag, Basel, Switzerland in 2002

Member of the BertelsmannSpringer Publishing Group
Printed on acid-free paper produced from chlorine-free pulp. TCF ∞
Cover design: Markus Etterich, Basel
Cover illustration: Photomicrograph of a dura perivascular mast cell examined with electron microscopy (see p. 116)

ISBN 3-7643-6231-6

9 8 7 6 5 4 3 2 1

www.birkhauser.ch

Contents

List of contributors vii

Preface ix

Egilius L.H. Spierings
Inflammation in migraine pathogenesis: when, where, and how 1

Ann I. Scher, Richard B. Lipton and Walter F. Stewart
Impact of migraine on a personal and societal level 21

Elizabeth W. Loder
Presentation, pathogenesis, and treatment of migraine 37

David W. Dodick
Neurogenic inflammation in the pathogenesis of migraine 55

Uwe Reuter and Guy Arnold
Models of neurogenic inflammation as it relates to migraine 65

Dimos D. Mitsikostas
Mediators and their receptors involved in neurogenic inflammation 91

Theoharis C. Theoharides and Kristiana Kandere
Mast cell involvement in neurogenic inflammation 115

Egilius L.H. Spierings
Inhibition of neurogenic inflammation in abortive migraine treatment 133

Margarita Sánchez del Río
Inhibition of neurogenic inflammation in preventive migraine treatment 145

Margarita Sánchez del Río
Is migraine a neuroinflammatory disease? 161

Index 165

List of contributors

Guy Arnold, Humboldt Universität zu Berlin, Charité, Neurologische Klinik und Poliklinik, Schumannstrasse 20–21, 10098 Berlin, Germany;
e-mail: guy.arnold@charite.de

David W. Dodick, Mayo Clinic, Department of Neurology, 13400 East Shea Boulevard, Scottsdale, AZ 85259, USA; e-mail: dodick.david@mayo.edu

Kristiana Kandere, Tufts University School of Medicine, New England Medical Center, 136 Harrison Avenue, Boston, MA 02111, USA

Richard B. Lipton, IMR, 1200 High Ridge Road, Stamford, CT 06905, USA;
e-mail: rlipton@imrinc.com

Elizabeth W. Loder, Spaulding Rehabilitation Hospital, Pain Inpatient Unit, 125 Nashua Street, Boston, MA 02114-1198, USA; e-mail: eloder@partners.org

Dimos D. Mitsikostas, Athens Naval Hospital, Headache Clinic, 70 Dinokratous Street, 11521 Athens, Greece; e-mail: dmitsikostas@ath.forthnet.gr

Uwe Reuter, Humboldt Universität zu Berlin, Charité, Neurologische Klinik und Poliklinik, Schumannstrasse 20–21, 10098 Berlin, Germany;
e-mail: uwe.reuter@charite.de

Margarita Sánchez del Río, Headache Unit, Department of Neurology, Fundación Hospital Alcorcón, Juan Carlos I University, Madrid, Spain;
e-mail: msanchezdelrio@hotmail.com

Ann I. Scher, National Institutes of Health, Neuroepidemiology Branch/NINDS, 10 Center Drive, 55220, Bethesda, MD 20892, USA; e-mail: schera@ninds.nih.gov

Egilius L.H. Spierings, Brigham and Women's Hospital, Harvard Medical School, 75 Francis Street, Boston, MA 02115, USA; e-mail: spierings@mediaone.net ter systems free to execute the pain response. This can be achieved by prejunctional inhibition of transmitter release more readily than by a polypharmacological approach to achieve postjunctional blockade of multiple mediators. Research for specific trigeminal terminal receptors that could be suitable medication targets, thus, remains intense.

Research into migraine pathogenesis has benefited enormously from true synergy between astute clinical observations and investigations and the development of antimigraine medications with highly specific and well-defined pharmacology. The main challenges that now face the field are the discovery of the underlying factors that predispose migraineurs to the migraine trigger factors and the relationship between neural central nervous system dysfunction and the activation of pain-producing trigeminovascular structures. Such understanding may lead to new approaches that improve the success rate of current migraine preventive strategies and these are clearly needed.

Richard J. Hargreaves, Ph.D.
Senior Director, Pharmacology and Imaging Research
Merck Research Laboratories, West Point, Pennsylvania

Acknowledgement
The editors would like to thank Richard J. Hargreaves, Ph.D., and the series editor, Michael J. Parnham, Ph.D., for critically reviewing the manuscripts.

Inflammation in migraine pathogenesis: when, where, and how

Egilius L.H. Spierings

Brigham and Women's Hospital, Harvard Medical School, 75 Francis Street, Boston, MA 02115, USA

Introduction

Inflammation is one of the three mechanisms that are thought to be involved in the pathogenesis of the migraine headache and for which there is *clinical* experimental evidence. The other two mechanisms for which such evidence exists are *extra*cranial arterial vasodilation and decreased inhibition of central pain transmission. In so-called migraine with aura, the headache is preceded by transient focal neurological symptoms, generally referred to as aura symptoms. In the following, the pathogenesis of the migraine headache, as well as that of the migraine aura, will be discussed. The traditional (sequential) concept of the pathogenesis of the migraine attack will be reviewed and an alternative (parallel) concept presented. Finally, the mode of action of the specific abortive antimigraine medications, the triptans and ergots, will be discussed on the basis of available experimental evidence in man.

Pathogenesis of the migraine headache

Arterial vasodilation

In the 1930s, Graham and Wolff were the first to study the mechanism of *extra*cranial arterial vasodilation in the pathogenesis of the migraine headache [1]. They observed that pressure exerted on the extracranial arteries temporarily decreased the intensity of the pain. In addition, they found that administration of ergotamine resulted in a decrease in intensity of the headache, parallel to a decrease in pulsation amplitude of the extracranial arteries. It was also observed that increasing the pressure of the cerebrospinal fluid by intrathecal injection of saline, thereby decreasing the pulsation amplitude of the *intra*cranial arteries, did not decrease the intensity of the pain [2]. This suggests that the intracranial arteries, cerebral or extracerebral, that is, dural or meningeal, do not significantly contribute to the pain of the migraine headache. It does not mean, however, that the large cerebral and meningeal arteries are not sensitive to pain [3].

Migraine: A Neuroinflammatory Disease?, edited by Egilius L.H. Spierings and Margarita Sánchez del Río

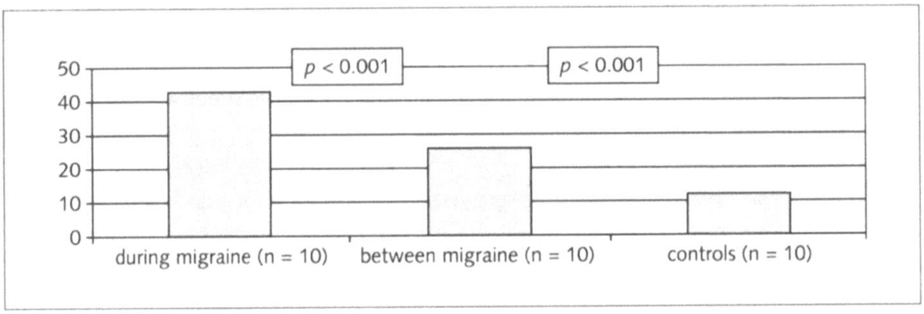

Figure 1
Systolic pulse-wave amplitude of the frontal branch of the superficial temporal artery, in mm, during and between migraine headaches and in non-headache controls. Data obtained from [4].

The artery that seems to be predominantly involved in the mechanism of migrainous vasodilation is the frontal branch of the superficial temporal artery, giving rise to the pain in the temple that is so characteristic of migraine. In 1953, Tunis and Wolff reported on the systolic pulse-wave amplitude of the frontal branch of the superficial temporal artery in migraineurs, during and between headaches, and in non-headache controls (Fig. 1) [4]. They found the amplitude, taken as a measure of artery caliber, to be significantly increased in between headaches in comparison to controls, with a further increase during headache.

More recent clinical experimental evidence for the involvement of the temporal artery in the pathogenesis of the migraine headache is shown in Figure 2 [5]. It shows that during the migraine headache, the artery is dilated on the side of the pain. The dilation is relative because generalized vasoconstriction occurs due to the activation of the sympathetic nervous system, secondary to the pain of the migraine headache. The generalized vasoconstriction causes the paleness of the face and coldness of the hands and feet, as is often seen during migraine headache. The fact that patients with migraine have dilated temporal arteries also between headaches was recently confirmed as well, as shown in Figure 3 [6].

Neurogenic inflammation

Neurogenic inflammation is inflammation of peripheral tissue, caused by the release of chemicals from the primary sensory nerve fibers involved in pain transmission. It finds its origin in the neurohumoral features of afferent nerve fibers, which were revealed in studies of the axon reflex flare (Lewis) and antidromic vasodilation (Baylis) [7]. The chemicals, which include substance P, calcitonin gene-related pep-

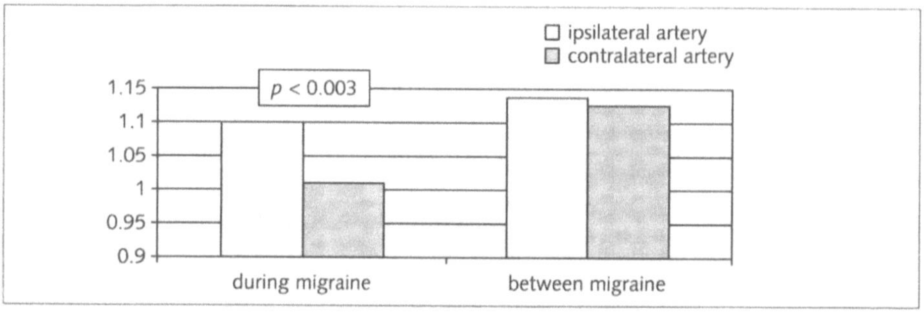

Figure 2
Luminal diameter of the superficial temporal artery, in mm, during and between migraine, ipsilateral and contralateral to the headache (n = 25). Data obtained from [5].

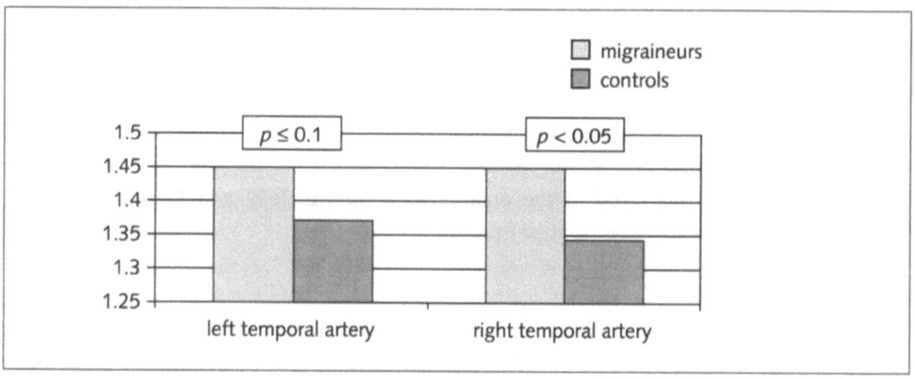

Figure 3
Luminal diameter of the superficial temporal artery, in mm, in migraineurs between headaches (n = 50), in comparison to non-headache controls (n = 50). Data obtained from [6].

tide and neurokinin A, are released from the nerve fibers when the fibers are activated. In migraine, the activation of the nerve fibers was thought to result from the dilation of the extracranial arteries. The nerve fibers coil around the arteries and are stretched, thereby depolarized and activated, when the blood vessels dilate. It was implicated in migraine to explain the decrease in pain threshold found locally at the site of the pain (Fig. 4) [8], and because it was understood that vasodilation alone could not account for the intensity of the migraine headache [9].

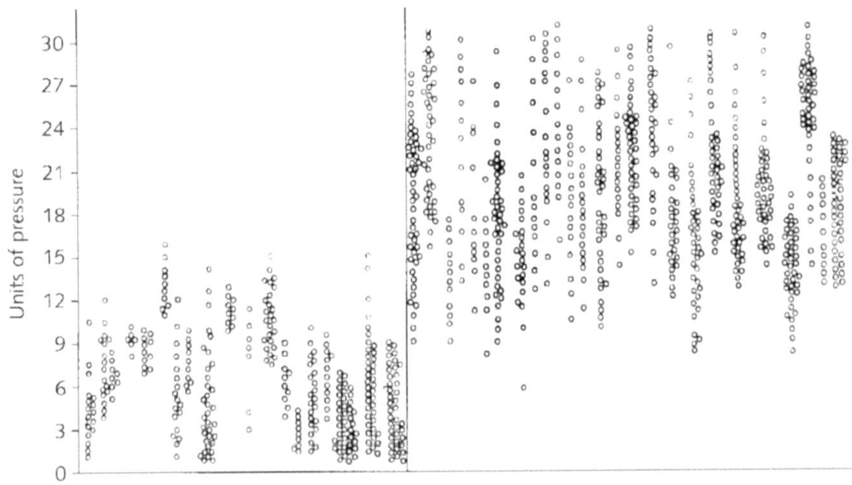

Figure 4
Deep pain threshold measured in the area of pain during (left) and between migraine headaches (right) (n = 10). Reproduced from [8].

In the 1950s, Chapman and Wolff were the first to study neurogenic inflammation as a mechanism involved in the pathogenesis of the migraine headache [10]. They observed that subcutaneous perfusates of sites of migraine headache possessed inflammatory activity proportional to the intensity of the pain (Fig. 5). In addition, they found that administration of ergotamine resulted in a decrease in inflammatory activity, parallel to a decrease in intensity of the pain. More recently, it was shown that during migraine headache, the level of calcitonin gene-related peptide in the external jugular vein is increased, in comparison to the antecubital vein (Fig. 6) [11]. Calcitonin gene-related peptide, a potent vasodilator, is one of the chemicals involved in neurogenic inflammation; the external jugular vein drains blood from the *extra*cranial tissues.

The decrease in pain threshold during migraine headache at the site of the pain was recently confirmed as cutaneous allodynia [12]. Allodynia refers to pain resulting from non-noxious stimulation of normal (looking) tissue and was found in the ipsilateral periorbital area during migraine headache in 79% of the 42 patients studied.

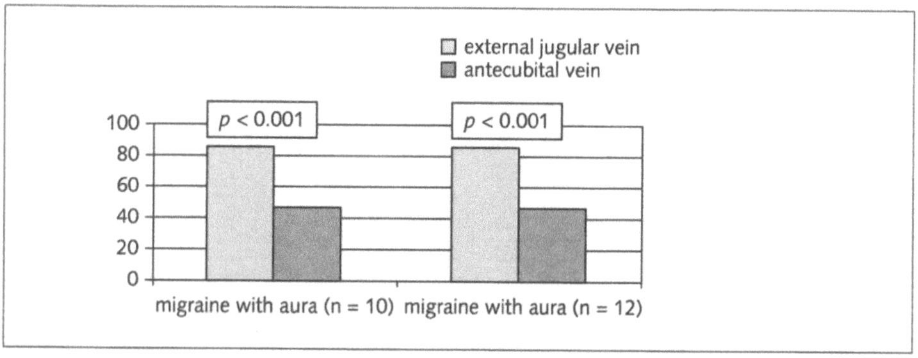

Figure 5
Inflammatory activity of subcutaneous perfusates of sites of migraine headache, in bradykinin units, in relation to the intensity of the pain. Data obtained from [10].

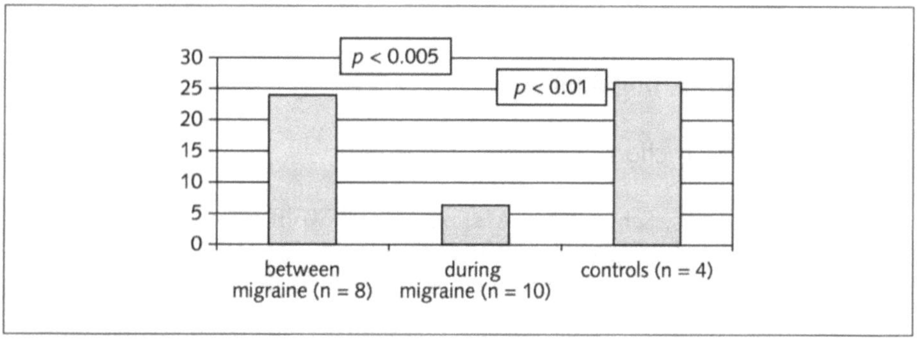

Figure 6
Level of calcitonin gene-related peptide in blood drawn from the external jugular vein, in pmol/l, during migraine headache in comparison to blood drawn from the antecubital vein. Data obtained from [11].

Central pain transmission

Apart from neurogenic inflammation, there is probably also a central mechanism involved in the decrease in pain threshold at the site of the migraine headache. Evidence for this was provided by a study of the enkephalin level, determined in the cerebrospinal fluid during migraine headache [13]. Enkephalin is an endogenous opioid and inhibits the transmission of pain signals in the central nervous system. It was found to be decreased during migraine headache in comparison to between headaches and to non-headache controls (Fig. 7).

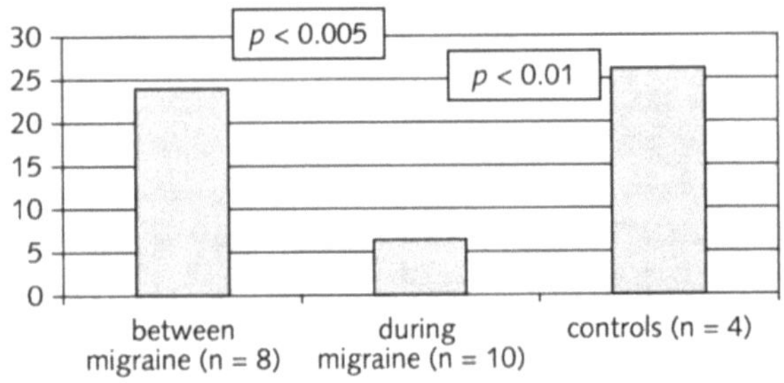

Figure 7
Enkephalin level of the cerebrospinal fluid, in pmol equivalents of met-enkephalin/ml, during and between migraine headaches and in non-headache controls. Data obtained from [13].

Pathogenesis of the migraine aura

Cerebral vasoconstriction

In the 1940s and 50s, Schumacher, Marcussen and Wolff were the first to experimentally study the pathogenesis of the migraine aura [2, 14]. They observed that inhalation of a cerebral vasodilator, such as amyl nitrite or carbon dioxide, during the migraine aura resulted in a transient regression of the symptoms (Fig. 8). Hence, they concluded that the migraine aura is caused by transient cerebral vasoconstriction.

Spreading depression

In 1958, Milner reported on the similarities in features and progression between the scintillating scotoma and spreading depression, a neurophysiological phenomenon described by Leão in 1944 [16]. Spreading depression is a wave of inhibition of neuronal activity, which travels over the cerebral cortex at a slow rate. It is preceded by a short-lasting phase of intense neuronal activity and, therefore, is better referred to as "spreading excitation". Scintillating scotoma is a typical presentation of the migraine aura. It usually begins near the center of vision as a twinkling star, which develops into a circle of bright and sometimes colorful, flickering zigzag lines. The circle subsequently opens up on the inside to form a semicircle or horseshoe, which further expands into the periphery of one visual field or the other. On the inside of the visual disturbance, a band of dimness follows in the wake of the

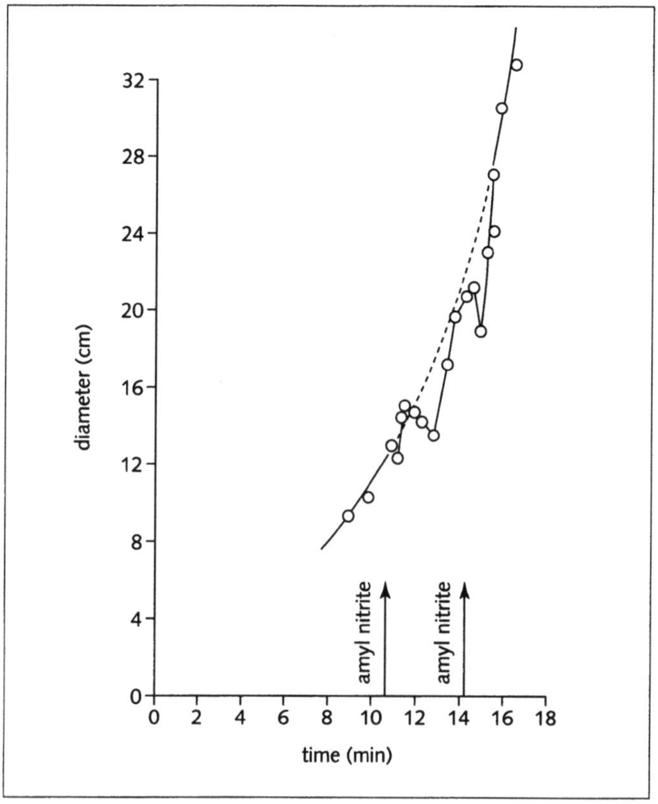

Figure 8
Effect of the cerebral vasodilator, amyl nitrite, on the progression of the scintillating scotoma. Reproduced from [15].

crescent of flickering zigzag lines. The disturbance of vision ultimately disappears as it fades away in, or moves outside of, the periphery of the visual field in which it developed.

Cerebral blood flow

In the 1970s, relatively accurate measurement of cerebral blood flow became possible with the development of the Xenon-clearance technique. Olesen et al. summarized the results of blood flow studies applying this technique in 63 patients with attacks of migraine with aura triggered by angiography [17]. They concluded that the aura symptoms come on *after* a decrease in blood flow occurs in the posterior

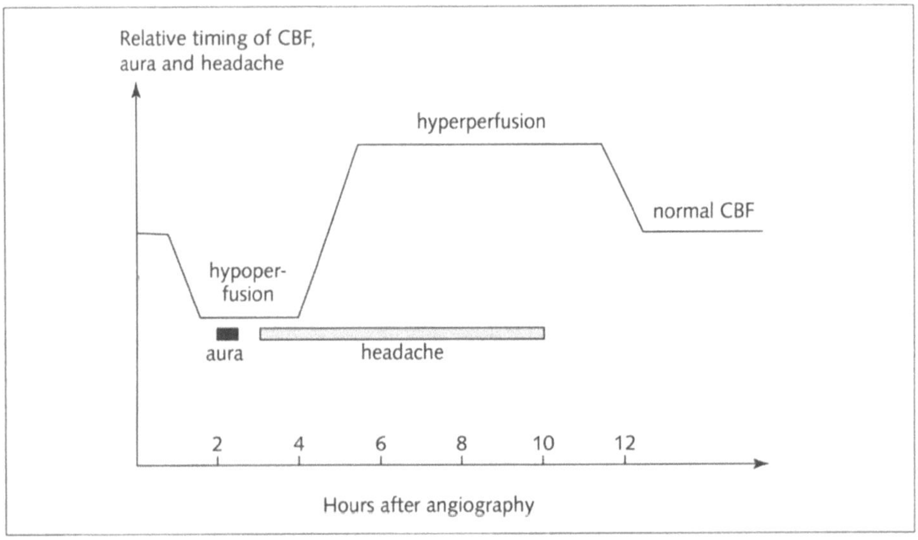

Figure 9
Changes in cerebral blood flow in relation to the occurrence of the aura and headache in migraine with aura. Reproduced from [17].

region of the opposite hemisphere. The headache comes on *while* blood flow is still decreased but is followed by a gradual increase in blood flow to an abnormally high level (Fig. 9). The increase in cerebral blood flow that follows the decrease was initially attributed to reactive hyperemia. However, the decrease did not reach ischemic levels, as is required to cause reactive hyperemia, and was, therefore, referred to as oligemia [18]. The oligemia was shown to spread over the cerebral cortex at a slow rate, similar to Leão's spreading excitation/depression.

The development of functional magnetic resonance imaging made it possible to study the changes in cerebral blood flow during spontaneous migraine attacks. The resolution of this method is also much better than that of the Xenon-clearance technique and, in addition, the brain can be studied in different planes. It was found that during the migraine aura, cerebral blood flow is decreased by 27% in the contralateral occipital cortex and this decrease persists for up to 2.5 h into the headache [19]. Whether cerebral blood flow subsequently increases, as suggested by the Xenon-clearance studies, is not clear. An increase of 20% over multiple attacks was observed in one of the three patients studied. In the patients with migraine without aura, no changes in cerebral blood flow were found 1–11 h after the onset of headache, as compared to between headaches.

A study using high-field functional magnetic resonance imaging looked at the blood oxygenation-level dependent signal during visual migraine aura [20]. The sig-

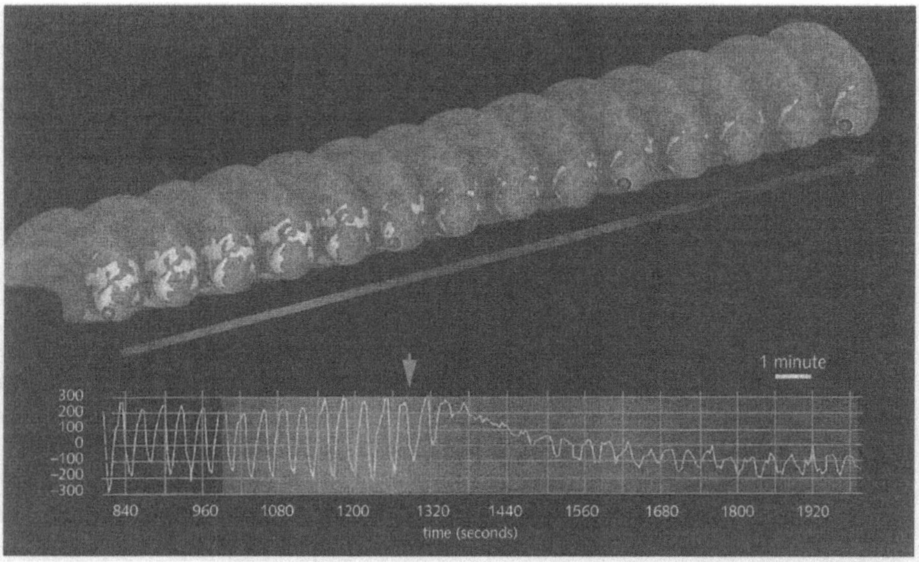

Figure 10
Above, high-field magnetic resonance images of an "inflated" right cerebral hemisphere taken over a time course of 20 min, including 12 min after the onset of exercise-induced migraine aura paracentrally in the left visual field (arrow). The circle projected on some of the images indicates the primary visual cortex or striate area. The activity shown is the so-called blood oxygenation-level dependant signal, elicited by visual stimulation through a flickering checkerboard and reflects the balance between oxygen delivery and oxygen consumption. At the onset of the migraine aura, the signal is suppressed starting in the striate area and gradually extending anteriorly at a rate of 3.5 mm/min. Below, a graphic display over the same time period of the amplitude of the blood oxygenation-level dependant signal. At the onset of the aura, the mean level of the amplitude increases markedly, which suggests heightened cortical neuronal activity. The increase in mean level is rapidly followed by almost complete abolition of the signal, indicating suppression of cortical neuronal activity despite continuing checkerboard stimulation. Courtesy of Margarita Sánchez del Río, M.D., Madrid, Spain; recorded at the Massachusetts General Hospital, Harvard Medical School, Boston, Massachusetts.

nal was elicited by visual stimulation through a flickering checkerboard and reflects the balance between oxygen delivery and oxygen consumption. Five attacks of migraine with aura were studied in three subjects, two of which were triggered in one subject by strenuous exercise. In particular in the triggered attacks, a sequence of events was recorded, consisting of a marked increase in the mean level of the signal, rapidly followed by its almost complete abolition despite continuing checkerboard stimulation (Fig. 10). The observed changes in the signal suggest a particular sequence of events, consisting of brief excitation followed by prolonged (active) depression of cortical neuronal activity. The changes started in the primary visual cortex or striate area and extended anteriorly over the occipital lobe at a rate of 3.5 mm/min, which is similar to the rate of progression of Leão's spreading excitation/depression.

Summary

On the basis of the clinical presentation of the migraine aura, the mechanism involved is more likely to be spreading excitation/depression than transient cerebral vasoconstriction. This notion is supported by the studies reviewed above, as well as by a study using brain spectroscopy, which showed an alteration in energy metabolism during migraine with aura but without changes in pH [21]. Cerebral vasoconstriction is the mechanism most likely involved in migrainous infarction, which can complicate a migraine headache whether or not it is preceded or accompanied by aura.

Pathogenesis of the migraine attack

Traditional (sequential) concept

It is important to remember that only in the minority of cases is the migraine headache preceded by an aura. Otherwise, we are dealing with migraine without aura, in which the headache occurs without aura but is otherwise the same. In the traditional view, the pathogenesis of the migraine aura and headache are considered to be causally related, that is, the aura is considered to be the *cause* of the headache. The aura is related to cerebral vasoconstriction, which causes localized hypoxia of the brain and is followed by reactive vasodilation. The vasodilation occurs in the cerebral circulation but is *supposedly* associated with dilation of extracranial arteries. The extracranial arterial vasodilation initiates the mechanism of neurogenic inflammation and the interplay of the two causes the pain of the migraine headache (Fig. 11).

In migraine without aura, as the traditional view maintains, the cerebral vasoconstriction and hypoxia occur as well but in a clinically silent area of the cerebral cortex. However, there is little evidence for this assumption and there is also no evidence that cerebral vasodilation is associated with dilation of extracranial arteries. The two assumptions were made to causally connect the migraine aura with the headache and to bring the two forms of migraine, migraine with and without aura, together in one pathogenetic concept. However, on the basis of the results of the

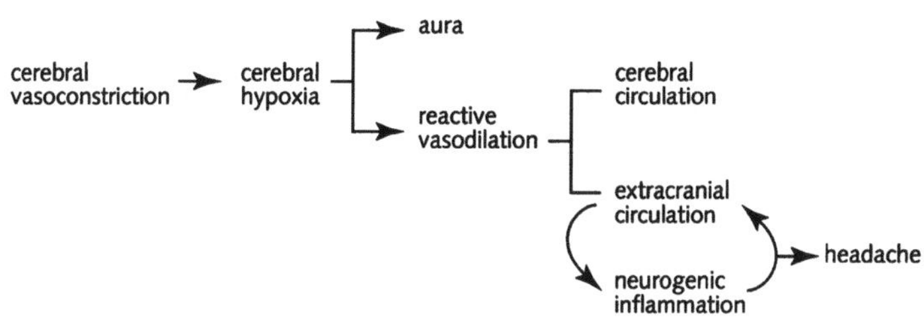

Figure 11
Traditional view on the pathogenesis of the migraine attack, in which the aura and headache are considered to be sequential and causally related.

cerebral blood flow studies, it can at least be stated that the aura and headache are not causally related through cerebral vasodilation because the cerebral vasodilation, if it happens at all, does not occur until after the onset of the headache.

Alternative (parallel) concept

Except for the aura, the clinical presentations of migraine with and without aura are so similar that a common pathogenesis is plausible. Also, the two forms of migraine often occur in the same individual, with some headaches preceded, and some not, by an aura. The fact that the aura often occurs before the onset of the headache does not necessarily make it the cause of the headache. It is relatively simplistic reasoning to assume that because one event follows the other, there is a causal relationship between the two.

The particular time-relationship between the occurrence of the aura and headache can also be explained in other ways. For example, with a disturbance in physiology, the reactivity of the cerebral cortex is much greater than that of the extracranial tissues in giving rise to symptoms. In the alternative concept, the pathogenesis of the migraine aura and headache are considered to be parallel rather than sequential in nature (Fig. 12). They are joined together by the migraine process, the driving force behind the migraine attack. The migraine process is activated, often in unison, by the migraine triggers. There is evidence from a psychophysiological study that the visual cortex of patients with migraine with aura lacks inhibitory activity, in comparison to patients with migraine without aura and non-headache controls [23]. This lack of inhibitory activity could translate into hyperexcitability or a lower

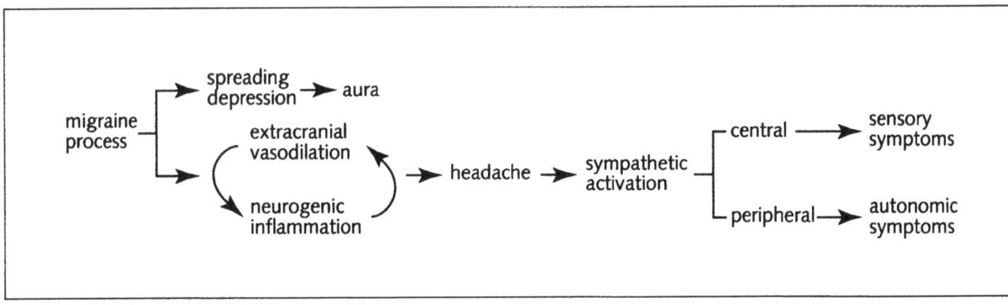

Figure 12
Alternative view on the pathogenesis of the migraine attack, in which the aura and headache are considered parallel phenomena and the associated symptoms are looked upon as secondary to the headache. Adapted from [22].

threshold for initiation of the spreading excitation/depression phenomenon by the migraine process. It is possible that this hyperexcitability, in turn, relates to a genetically determined calcium channelopathy in the central nervous system [24].

The parallel concept explains better than the sequential one the isolated occurrence of the migraine aura (migraine aura without headache) and the isolated occurrence of the migraine headache (migraine without aura). The concept also includes the associated, autonomic and sensory symptoms of the migraine headache, explained as secondary to the headache through stimulation of the sympathetic nervous system and ascending reticular activating system, respectively.

Migraine generator

A study using positron emission tomography showed cerebral blood flow to be increased during the migraine headache in the cingulate as well as visual and auditory cortex [25]. The activation of the cingulate cortex, part of the limbic system, is directly related to the pain of the migraine headache and the emotional response to it. The activation of the visual and auditory associate cortex has been related tentatively to the photophobia and phonophobia. In addition to these changes in cortical cerebral blood flow, an increase in blood flow was observed in the contralateral brainstem at the level of the mesencephalon (Fig. 13). This change in blood flow, reflecting increased neuronal activity, was thought to be particular to migraine because it did not occur with capsaicin-induced head pain [26]. It was also found that while administration of sumatriptan resulted in relief of the migraine symptoms and disappearance of the cortical activation, the localized increase in mesencephalic blood flow persisted. It was, therefore, suggested that the observed activation in

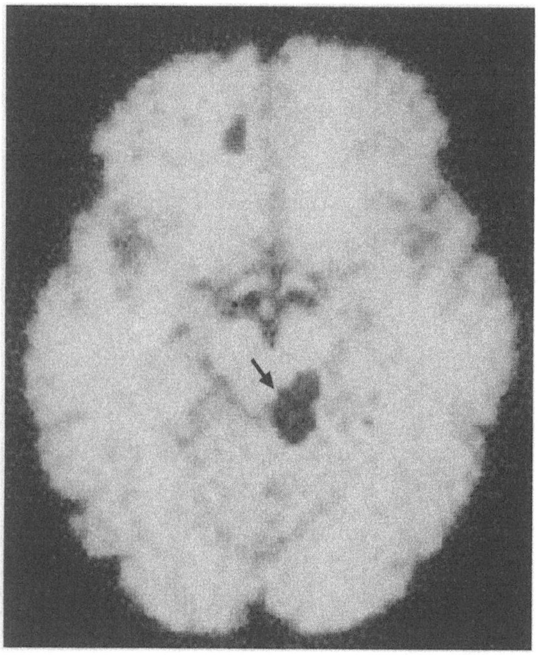

Figure 13
Increased blood flow in an area of mesencephalon contralateral to the pain in patients with unilateral migraine headache (arrow), as observed with positron emission tomography (n = 9). Also shown is the increased blood flow in the ipsilateral cingulate cortex. Reproduced from [25].

the brainstem contralateral to the pain was the first visualization of a postulated migraine generator. From this generator, the above-mentioned migraine process could be initiated, activated by the migraine trigger factors.

Unfortunately, the finding of localized brainstem activation during the migraine headache has not been replicated yet, which would be the first step in establishing its significance. Also, the brainstem area involved is thought to play an important role in central pain control rather than in the generation of pain. Painful thermal stimulation (50 °C) of the arm has been shown to increase blood flow in the same mesencephalic area, contralateral to the side of stimulation (Fig. 14) [27]. Furthermore, it is not known how activation of this area could cause the cerebrocortical and extracranial changes related to the migraine aura and headache, respectively. Equally obscure are the mechanisms and pathways through which this brainstem area could be activated by the diversity of trigger factors involved in migraine.

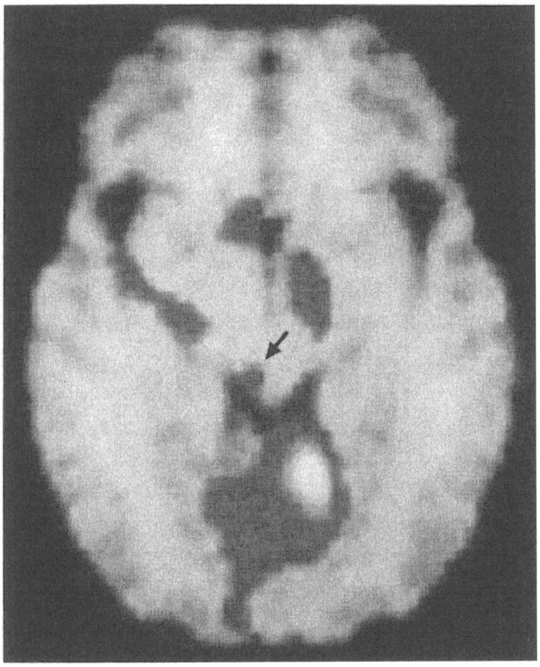

Figure 14
Increased blood flow in the contralateral mesencephalon in the area of the periaqueductal gray (arrow), as observed with positron emission tomography during painful thermal stimulation (50 °C) of the arm (n = 10). Reproduced from [27].

Pharmacology of migraine abortion

Modes of action of the ergots

Of the specific abortive antimigraine medications, ergotamine is the oldest and was introduced in the treatment of migraine in 1926. Its potent constrictor effect on the extracranial (temporal and occipital) arteries in man is shown in Figure 15. Ergotamine has an effect on neurogenic inflammation as well, which is illustrated in Figure 16 in relation to its effect on the intensity of the migraine headache.

A derivative of ergotamine, dihydroergotamine, was introduced in 1945 in the abortive treatment of migraine. Its pharmacology is very similar to that of ergotamine with quantitative rather than qualitative differences. It is four times less potent as a vasoconstrictor but ten times less potent as an emetic agent and is, therefore, better tolerated with fewer gastrointestinal side-effects. The potent vasoconstrictor effect of dihydroergotamine on the superficial temporal artery in man is shown in Figure 17.

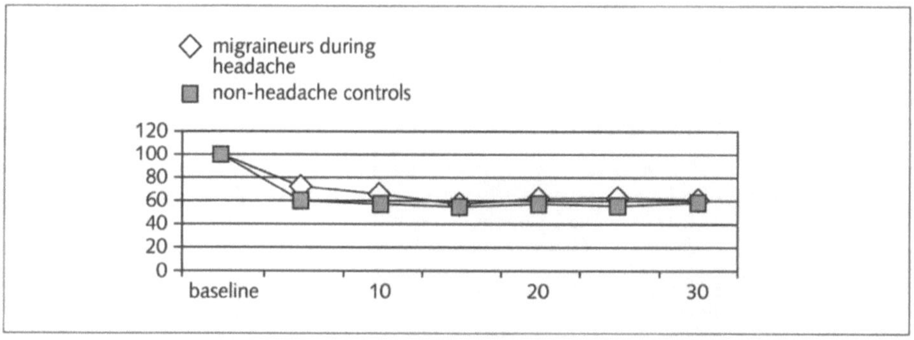

Figure 15
Percentage decrease in pulsation amplitude of the extracranial (temporal and occipital) arteries during migraine headache (n = 20) and in non-headache controls (n = 34), after intravenous administration of 0.37–0.5 mg ergotamine. Data obtained from [1].

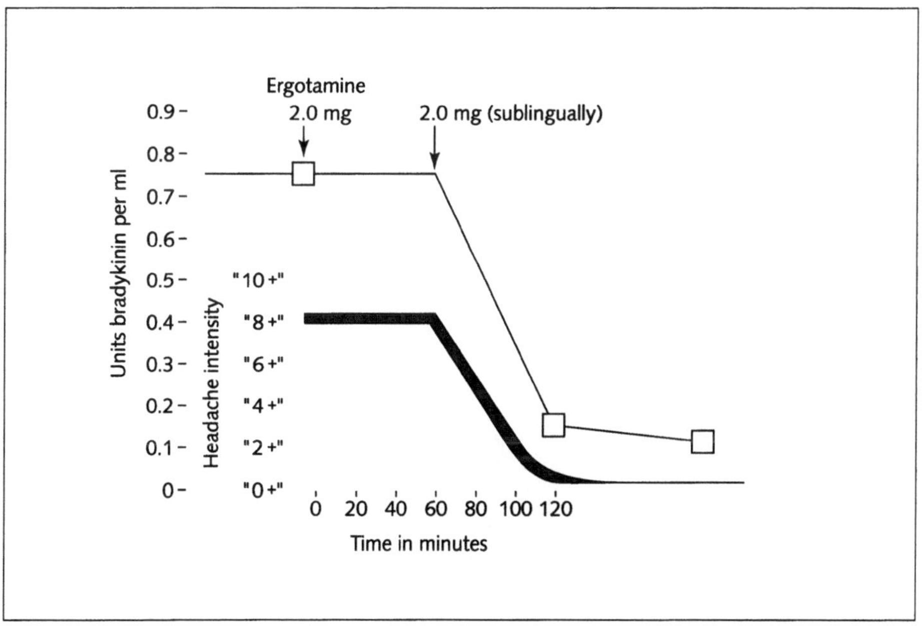

Figure 16
Effect of ergotamine, 4 mg sublingually, on the inflammatory activity of subcutaneous perfusates of sites of migraine headache, in bradykinin units, in relation to the effect on the intensity of the pain. Reproduced from [10].

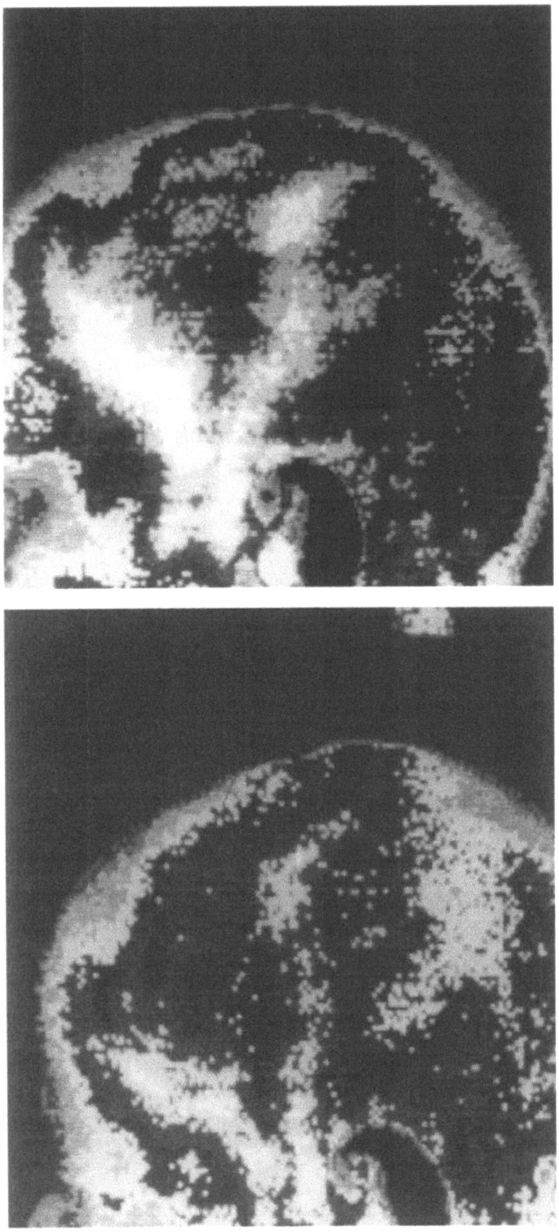

Figure 17
Thermograms of the lateral side of the head, showing the superficial temporal artery in white before (above) and 30 min after intravenous administration of 0.5 mg dihydroergotamine (below).

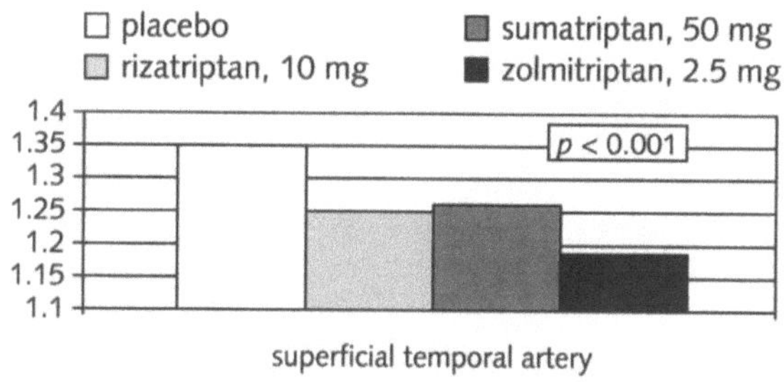

Figure 18
Effect of oral triptans, 1.5–2.5 h after administration, in comparison to placebo, on the luminal diameter of the superficial temporal artery, in mm, between migraine headaches (n = 16). Data obtained from [6].

Modes of action of the triptans

The triptans were developed in the 1980s and 90s on the basis of the vascular mode of action of the ergots. The first triptan that became available on the market was sumatriptan. Contrary to general belief, sumatriptan was *not* developed specifically for the treatment of migraine but as a pharmacological tool. It was developed as a selective agonist of the serotonin 1-like receptor that mediates contraction of the dog saphenous vein, in order to distinguish it from the serotonin 1-like receptor that mediates relaxation of the cat saphenous vein [28]. The selective serotonin 1-like receptor agonist, 5-carboxamide tryptamine, had been shown in animals to redistribute common carotid blood flow in the same way, as had been demonstrated for the ergots. The redistribution of carotid blood flow by the ergots was not due to the reduction they cause in carotid blood flow but to their potent constrictor effect on arteriovenous anastomoses [29]. On the basis of measurements of the oxygen saturation of blood drawn from the external jugular vein and an artery, it had been postulated that dihydroergotamine aborts migraine headaches by constricting arteriovenous anastomoses [30]. This is what established the link between the serotonin 1-like receptor, particularly the one mediating contraction of the isolated dog saphenous vein, and migraine. It was this link, rather than the observations made in the 1960s that linked serotonin and migraine, which led to the introduction of sumatriptan and, thereby, of the triptans in the abortive treatment of migraine.

The triptans, as is shown in Figure 18 for rizatriptan, sumatriptan and zolmitriptan, constrict the superficial temporal artery in man. In addition, they have been

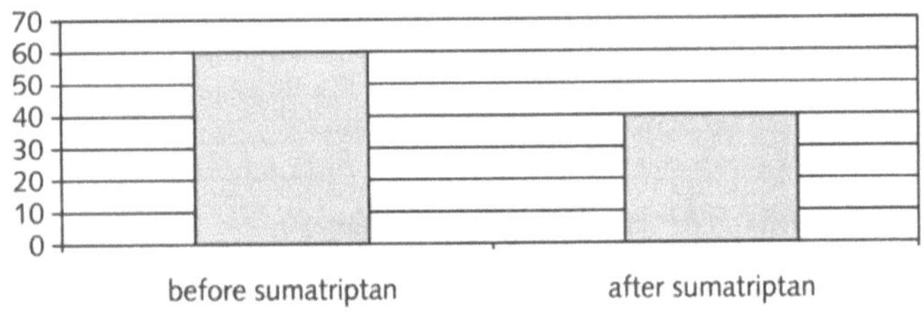

Figure 19
Effect of sumatriptan, 3 or 6 mg subcutaneously, on the level of calcitonin gene-related peptide in blood drawn from the external jugular vein, in pmol/l, during migraine headache (n = 8). Data obtained from [31].

demonstrated to inhibit the release of calcitonin gene-related peptide, the neuropeptide that mediates the vasodilation component of neurogenic inflammation, as shown in Figure 19 for sumatriptan.

Conclusion

There is clinical evidence that neurogenic inflammation is involved in the pathogenesis of the migraine headache. The inflammation results in a lowering of the superficial and deep pain thresholds. It was postulated to be secondary to the vasodilation, accentuating it and making it extremely painful. There is clinical evidence for at least the involvement of calcitonin gene-related peptide in mediating the inflammation. Finally, the selective abortive antimigraine medications, the ergots and triptans, not only cause vasoconstriction but also reduce neurogenic inflammation, as shown in man.

References

1 Graham JR, Wolff HG (1938) Mechanism of migraine headache and action of ergotamine tartrate. *Arch Neurol Psychiat* 39: 737–763
2 Schumacher GA, Wolff HG (1941) A. Contrast of histamine headache with the headache of migraine and that associated with hypertension. B. Contrast of vascular mechanisms in preheadache and in headache phenomena of migraine. *Arch Neurol Psychiat* 45: 199–214

3 Ray BS, Wolff HG (1940) Pain-sensitive structures of the head and their significance in headache. *Arch Surg* 41: 813–856

4 Tunis MM, Wolff HG (1953) Long-term observations of the reactivity of the cranial arteries in subjects with vascular headache of the migraine type. *Arch Neurol Psychiat* 70: 551–557

5 Iversen HK, Nielsen TH, Olesen J, Tfelt-Hansen P (1990) Arterial responses during migraine headache. *Lancet* 336: 837–839

6 De Hoon JNJM (2000) *Migraine and anti-migraine drugs. Focus on cardiovascular aspects.* Doctoral Thesis, Maastricht University, Maastricht, The Netherlands

7 Chapman LF, Ramos AO, Goodell H, Wolff HG (1961) Neurohumoral features of afferent fibers in man. Their role in vasodilatation, inflammation, and pain. *Arch Neurol* 4: 49–82

8 Wolff HG, Tunis MM, Goodell H (1953) Evidence of tissue damage and changes in pain sensitivity in subjects with vascular headache of the migraine type. *Arch Int Med* 92: 478–484

9 Dalessio DJ (1972) *Wolff's headache and other head pain.* Oxford University Press, New York, 272–307

10 Chapman LF, Ramos AO, Goodell H et al (1960) A humoral agent implicated in vascular headache of the migraine type. *Arch Neurol* 3: 223–229

11 Goadsby PJ, Edvinsson L, Ekman R (1990) Vasoactive peptide release in the extracerebral circulation of human during migraine headache. *Ann Neurol* 28: 183–187

12 Burnstein R, Yarnitsky D, Goor-Areyh I et al (2000) An association between migraine and cutaneous allodynia. *Ann Neurol* 47: 614–624

13 Anselmi B, Baldi E, Casacci F, Salmon S (1980) Endogenous opioids in cerebrospinal fluid and blood in idiopathic headache sufferers. *Headache* 20: 294–299

14 Marcussen RM, Wolff HG (1950) 1. Effects of carbon dioxide-oxygen mixtures given during preheadache phase of the migraine attack. 2. Further analysis of the pain mechanisms in headache. *Arch Neurol Psychiat* 63: 42–51

15 Hare EH (1966) Personal observations on the spectral march of migraine. *J Neurol Sci* 3: 259–264

16 Milner PM (1958) Note on a possible correspondence between the scotomas of migraine and spreading depression of Leão. *Electroencephalogr Clin Neurophysiol* 10: 705

17 Olesen J, Friberg L, Olsen TS et al (1990) Timing and topography of cerebral blood flow, aura, and headache during migraine attacks. *Ann Neurol* 28: 791–798

18 Olesen J, Larsen B, Lauritzen M (1981) Focal hyperemia followed by spreading oligemia and impaired activation of rCBF in classic migraine. *Ann Neurol* 9: 344–352

19 Sanchez del Rio M, Bakker D, Wu O et al (1999) Perfusion weighted imaging during migraine: spontaneous visual aura and headache. *Cephalalgia* 19: 701–707

20 Hadjikhani N, Sanchez del Rio M, Wu O et al (2001) Mechanisms of migraine aura revealed by fMRI in human visual cortex. *Proc Natl Acad Sci USA* 98: 4687–4692

21 Welch KMA, Levine SR, D'Andrea G et al (1989) Preliminary observations on brain

energy metabolism in migraine studied by *in vivo* phosphorus 31 NMR spectroscopy. *Neurology* 39: 538–541
22 Spierings ELH (1988) Recent advances in the understanding of migraine. *Headache* 28: 655–658
23 Palmer JE, Chronicle EP, Rolan P, Mulleners WM (2000) Cortical hyperexcitability is cortical under-inhibition: evidence from a novel functional test of migraine patients. *Cephalalgia* 20: 525–532
24 Ophoff RA, Terwindt GM, Vergouwe MN et al (1996) Familial hemiplegic migraine and episodic ataxia type-2 are caused by mutations in the Ca^{2+} channel gene CACNL1A4. *Cell* 87: 543–552
25 Weiller C, May A, Limmroth V et al (1995) Brain stem activation in spontaneous human migraine attacks. *Nature Med* 1: 658–660
26 May A, Kaube H, Buchel C et al (1998) Experimental cranial pain elicited by capsaicin: a PET study. *Pain* 74: 61–66
27 Paulson PE, Minoshima S, Morrow TJ, Casey KL (1998) Gender differences in pain perception and patterns of cerebral activation during noxious heat stimulation in humans. *Pain* 76: 223–229
28 Spierings ELH (2000) The (suma)triptan history revisited (letter). *Headache* 40: 766–767
29 Spierings ELH, Saxena PR (1979) Effect of ergotamine on cranial arteriovenous shunting in experiments with constant flow perfusion. *Eur J Pharmacol* 56: 31–37
30 Heyck H (1969) Pathogenesis of migraine. *Res Clin Stud Headache* 2: 1–28
31 Goadsby PJ, Edvinsson L (1991) Sumatriptan reverses the changes in calcitonin gene-related peptide seen in the headache phase of migraine. *Cephalalgia* 11 (Suppl 11): 3–4

Impact of migraine on a personal and societal level

[1]Ann I. Scher, [2]Richard B. Lipton and Walter F. Stewart

[1]National Institutes of Health, Neuroepidemiology Branch/NINDS, 10 Center Drive, 55220, Bethesda, MD 20892, USA; [2]Richard B. Lipton, IMR, 1200 High Ridge Road, Stamford, CT 06905, USA

Introduction

Migraine is a common, temporarily disabling, episodic disorder that affects about 28 million adults in the United States [1]. The World Health Organization classified a severe migraine attack requiring bed rest as one of the most disabling of the 30 diseases considered [2]. However, even between attacks, migraineurs live in fear that the next attack could disrupt their work, social, or family responsibilities. Thus, there is likely to be some large, but unstudied, chronic disability associated with migraine even between attacks.

The burden of an illness to society consists of both economic and social costs. Economic costs include direct costs of treatment (medications, physician visits, and hospitalizations), as well as indirect costs due to disability (absenteeism and reduced productivity). Social costs include the more intangible aspects of illness, such as reduced quality of life, family impact, and pain. As physician consultation is not the rule for those suffering from migraine, the direct cost of migraine treatment is largely borne by the individual. The greater economic burden of migraine is due to missed work and reduced productivity and, therefore, is concentrated on the employed adult population. As migraine is a common disorder that tends to occur most frequently during the peak productive years, the quantifiable costs to society are substantial. We will first review the epidemiology of migraine and focus on the demographic factors that are associated with migraine prevalence. We will then focus on the economic and social impact of the illness in North America and Europe.

Epidemiology

Over 90 studies have been published on the epidemiology of migraine since the 1960s [3, 4]. In the past, prevalence estimates for migraine have varied considerably, largely due to differences in diagnostic criteria and underlying demographic differ-

Migraine: A Neuroinflammatory Disease?, edited by Egilius L.H. Spierings and Margarita Sánchez del Río

ences in the populations studied. Recent estimates based on the diagnostic criteria as specified by the International Headache Society [5] have been fairly consistent, with about 11% of adults and 5–10% of children in North America and Europe classified as suffering from migraine [3]. Migraine may be somewhat less prevalent in South America [6–10] and Asia [11–16].

Variation by gender and age

Migraine is equally prevalent in men and women, or perhaps more prevalent in men, only until about the age of 10 [7, 17–21]. Incidence increases more rapidly during adolescence in girls than in boys, with the result that migraine is, on average, two to three times more common in adult women than in adult men. For example, the prevalence of migraine in the twenties through forties was about 21% for women and 10% for men in a study conducted in the United States (unpublished data from [22]). While attacks tend to become less severe and frequent for both men and women after the forties, this female preponderance persists even after the age of menopause [23].

Variation by race

Several surveys in the United States and United Kingdom conducted in multi-racial populations found somewhat higher migraine prevalence in whites than blacks [1, 22, 23] or Asians [22]. However, prevalence was similar by race in another study in the United States [24]. Results by race, in general, are compatible with meta-analytic results, suggesting that migraine is most common in North America and Western Europe and less common in Asia and Africa [3].

Frequency and transformed/chronic migraine

Migraine sufferers typically experience about one attack per month [25]. Each attack lasts about 24 h on average [25], although it should be noted that, by International Headache Society definition, attacks must last at least 4 h for at least five lifetime attacks. Women tend to have more frequent and longer-lasting attacks than men do [25].

While most individuals with migraine suffer clear-cut episodic attacks with pain-free intervals, there is an important subset, consisting of about 8% of the adult migraine population, which experiences daily or almost-daily headaches [30]. Clinically, this disorder is often referred to as "transformed migraine" and is defined by headaches that occur 15 or more days per month, with a duration of 4 h or greater on average [27]. Usually, there is a history of episodic migraine attacks that gradually increased in frequency, often while decreasing in intensity, to the point of daily or near-daily headache. The risk factors for transformed migraine are unknown and it may be that this condition – or daily headache in general – represents a progressive headache syndrome [28–30].

Co-morbidity

Individuals with migraine tend to have an increased frequency of co-morbid health conditions [31–35] and increased medical expenses for other health problems [32], compared to others of the same gender and age. Specific conditions that tend to co-occur more frequently than expected are epilepsy [36–39], stroke in women below the age of 45 [40, 41], anxiety, depression and mania [24, 31, 35, 42–49]. While depression is often found in those suffering from chronic pain [50–52], the association between migraine and depression appears to be bi-directional. That is, individuals with migraine are at increased risk of depression and individuals who are depressed are at increased risk of migraine [49, 53].

It should be noted that the increased medical costs incurred by migraineurs might be due not only to co-morbid conditions. It may also be due to a general increased likelihood of use of medical services or, perhaps, directly as a result of migraine therapy, for example, due to prolonged use of analgesics or to side-effects.

Social costs of migraine and reduced quality of life

Non-monetary costs to the individual in terms of pain, reduced quality of life, and lost opportunity are difficult to quantify. However, migraine sufferers often report negative effects on family relationships or family members [54–56]. Furthermore, unemployment is a risk among individuals with severe headaches: a total of 13% of patients who sought care for headache in a managed-care organization were unable to keep full-time employment in the prior year because of their headaches. The unemployment rate was about twice as high in the migraine patients than in those with non-migraine headache [51].

Consistent with this finding is that, in the United States, migraine prevalence is inversely related to income or educational level [22–24, 56]. This is the reverse of the pattern seen with tension-type headache, which tends to be more prevalent in those with higher socioeconomic status [57]. Disabling migraine may interfere with education and job performance leading to "downward drift". This phenomenon is sometimes called social causation, because illness causes a decline in socioeconomic status. It is also possible that those of lower income or educational level are more likely to have migraine because of poor access to medical care or greater exposure to environmental triggers of headache. This process is called "social selection". It should be noted that studies outside the United States have generally not found an association, either positive or negative, with migraine and measures of socioeconomic status, such as occupation, income, or education [17, 58–62]. Thus, there may be cultural or environmental factors unique to the United States that are associated with increased prevalence, intensity, or duration of migraine in those of lower socioeconomic status.

The impact of migraine on health-related quality of life has been studied using both generic and disease-specific instruments. The most widely used generic instrument is from the Medical Outcomes Study (Short Form-36) [63] and has been used in clinic- and population-based samples. In clinic-based studies, health-related quality of life in individuals with migraine who consulted physicians are compared to population norms, with clinic samples, or with individuals with other chronic conditions. Migraine sufferers identified in this way are likely to be more severely affected than those in the general population. As a consequence, health-related quality of life estimates from clinic samples are biased towards indicating a more severe impact of migraine on quality of life. The influence of selection bias on health-related quality-of-life estimates for migraine and for other conditions must also be considered, when comparing health-related quality-of-life estimates for migraine to other chronic conditions.

One such study was based on a group of patients at a headache clinic, consisting of both tertiary and primary care self-referrals within a multi-specialty group practice [64]. Health-related quality of life was measured using the Short Form-36 and was compared to previously published health scores from patients with other chronic conditions (depression, back pain, arthritis, myocardial infarction within the past year, congestive heart failure, symptomatic infection with human immunodeficiency virus). The authors concluded that the headache patients had health-related quality-of-life decrements similar to patients with myocardial infarction and congestive heart failure and a greater decrement than those associated with arthritis, diabetes, depression, and back pain [64]. The only chronic condition associated with a worse status was active infection with human immunodeficiency virus [64].

Another similar study was based on a sample of migraine sufferers who had previously participated in a placebo-controlled clinical trial [65]. Health-related quality of life scores were compared to the published scores of patients with other chronic conditions, such as hypertension, depression, osteoarthritis, and diabetes type II, which had been adjusted to reflect the age and gender distribution of the migraine patients. Unlike the first study, the health-related quality-of-life scores were adjusted to account for the presence of one or more co-morbid health conditions. Overall, the health-related quality-of-life scores for the migraine patients were roughly similar to patients with osteoarthritis or diabetes in terms of physical function and worse on scales measuring well-being.

Four population-based surveys compared health-related quality-of-life measures between individuals with and without migraine. In two studies conducted in The Netherlands, individuals with migraine had lower scores on most items in the European quality-of-life survey [32] or Short Form-36 [66], when compared to individuals without migraine in the general population. Quality-of-life surveys do not appear to differ between individuals who have migraine with *versus* migraine without aura. Similarly, two case-control studies conducted in the United States using the Short Form-36 [67], or the abridged Short Form-12 [68], found lower quality of life scores in migraine sufferers compared to population controls. The decreased health-related quality of life found in migraine was not explained by the increased prevalence of depression among the migraine sufferers: migraine and depression each exerted a significant and independent negative influence on health-related quality of life [68].

Economic costs of migraine

Economic burden is traditionally divided into the direct and indirect costs of illness. Direct costs include the costs of medical care such as physician consultation, medication, emergency room visits, and diagnostic testing. Indirect costs comprise the loss due to absenteeism and reduced productivity in paid employment and other roles. We consider each of these below.

Direct economic costs

Many of those with migraine are not currently being treated by a physician for headache and treat the condition exclusively with non-prescription medications. The direct cost of migraine treatment for this group, probably representing about half of the total migraine population, is borne by the individual and is not reflected in the cost analyses. However, non-prescription sales of pain medication (for all conditions) were estimated to be 3.2 billion dollars in the United States in 1999 [69] and headache accounts for about one-third of non-prescription analgesic use. Even for those who seek medical care, the costs incurred specifically for migraine are relatively modest, ranging from about 100 to 800 dollars per patient year in managed-care organizations (Tab. 1). It has been estimated that less than 10% of the total economic cost of migraine is borne by insurance, the rest being paid for by the patient (non-prescription medication, lost wages) or employer (absenteeism or reduced productivity) [70]. Due to the high prevalence of migraine in working-age populations, however, the overall cost to society is comparable to, or higher than, other chronic diseases, such as diabetes or asthma [70]. The specific direct costs of migraine treatment are summarized below.

Table 1 - Summary of studies on the annualized insurance direct costs of migraine treatment (United States)

Ref.	Expense	Amount	Comment
[84]	Physician/hospital	$127	Based on members of a large managed-care organization
	Excess physician/hospital compared to non-migraineurs	$381	Members with migraine were compared to matched members without migraine; migraineurs had relatively small expenses coded for migraine ($127) but had higher overall medical expenses
	Excess pharmacy compared to non-migraineurs	$158	
[85]	Physician/emergency room/ hospital/medication	$135 $324	Based on managed-care headache patients who had received sumatriptan
[70]	Physician/emergency room/ hospital/medication	$86–$97	Medstat MarketScan data; migraine-related claims from employees and dependants from 40 large employers
[86]	Physician/hospital/ emergency room/medication	$200–$800	Based on members of a large managed-care organization; medical expenses varied depending on the intensity of headache
[78]	Emergency room/clinic/hospital	$817	Clinical trial participants
[87]	Medical/radiology/medication	$357	Based on managed-care headache patients who had received sumatriptan

Physician consultation

The National Ambulatory Medical Care Survey estimated that there were over 10 million office or outpatient visits in the United States for headache in 1995 [71]. About half to two-thirds of those with migraine have consulted physicians specifically for headache at some time in their lives, with approximately half currently under physician's care (Tab. 2). Rates of current consultation have increased considerably in the United States over the last 10 years, from about 16% in 1989 to 48% in 1998 (unpublished data). Specialist consultation is uncommon in the United States (unpublished data) and Europe [32, 72]. Nonetheless, about 18% of office visits to neurologists in the United States in 1992 were for headache [73].

Prescription medications

Almost all migraineurs use some type of medication to treat their headaches, although only a minority uses prescription medication (Tab. 2). The most frequently used non-prescription medications are acetaminophen and ibuprofen. [55, 74] The most commonly used prescription medications are butalbital/aspirin/caffeine combinations, acetaminophen/codeine combinations, or triptans (unpublished data). Preventive medications are used by fewer than 10% of migraine sufferers [54, 55, 75].

Emergency room use and hospitalization

The use of emergency room services for migraine is surprisingly high, in the range of 13–24% in population samples (Tab. 2). According to the National Hospital Ambulatory Medical Care Summary, headaches were the fourth most common reason for visits to the emergency room (after stomach pain, chest pain, and fever) [76, 77]. Almost half of the participants in a clinical trial, presumably a more severely affected population, had visited an emergency room because of headache in the previous year [78]. Studies of patients who had a medical diagnosis of migraine also found somewhat higher rates of emergency room use (Tab. 2). These latter samples were limited to patients who had a documented medical consultation for migraine or headache and/or a migraine-specific pharmacy claim. Rates of hospitalization are 7% in the United States (clinical trial group) [78] and no more than 2% in Denmark [72].

Non-physician consultation and alternative medicine

Consultation with non-physician practitioners is fairly common in the United States and Europe. A recent study in the United States reported that 24% of those with severe headache had used alternative health care in the last year for headache, most commonly chiropractic manipulation [79]. In another recent survey in the United States [80], 32% of those reporting headaches had used alternative therapies within the last year and 13% had visited an alternative practitioner. The therapies most frequently used were relaxation therapy and chiropractic manipulation.

Similar results have been reported in Europe. For example, 17% of migraineurs had consulted an alternative practitioner within the last year in The Netherlands [32]. However, because those who consulted alternative practitioners had, on average, more visits than those who consulted general practitioners, a disproportionate share of all direct costs for migraine treatment (80%) was accounted for in this manner [32]. Other European studies have reported annual consultation rates for alternative practitioners ranging from 5–13% in Denmark [73] and England [55].

Table 2 - Summary of studies on physician consultation and use of prescription medication for migraine

Ref.	Country	Physician consultation		Prescription medications	Used emergency department	Comments
		Current (%)	Ever (%)	(%)	(%)	
[77]	US			28% (men) 40% (women)	13% (men) 20% (women)	
[54]	Canada		64%	44%	14%	
[74]	Canada			12%		
[56]	US			47%		
[1]	US		65%	41%	24%	Unpublished data from American Migraine Study II
Lipton (unpubl.)	US	48%	68%	45%		Migraineurs with at least 6 attacks per year
[72]			56%			
[13]	Japan	15%	31%	24%		
[24]	UK	49%	86%	31%		Migraineurs with at least 6 attacks per year
[22]	US		63%			
[88]	US	16%	63%	36%		
[32]	Netherlands	≈ 50%	≈ 70%			
[84]	US				18% (1-year)	Based on managed-care patients with migraine diagnosis and pharmacy claim
[85]	US				1.6% (6-months)	Based on managed-care headache patients who had received sumatriptan
[89]	US				average 3 visits/year	Based on managed-are headache patients who had received sumatriptan – average of before and after sumatriptan use

Table 2 - (continued)

Ref.	Country	Physician consultation		Prescription medications	Used emergency department	Comments
		Current (%)	Ever (%)	(%)	(%)	
[87]	US				16% (6-month period)	Based on managed-care headache patients who had received sumatriptan
[78]	US				48% (1-year)	
[55]	UK	40% (3 months)		28%		Hospital employees; 81% women
[90]	Spain	7%				Employees of 11 large companies
[75]	UK	19%	58%	10%		

Indirect economic costs

In the United Kingdom, 16% of all absences in one study of civil servants were due to headache or migraine [81]. On average, 2–3 workdays per migraineur per year are lost due to headache (Tab. 3), with the most disabled 20% accounting for 77% of the lost work days [82]. However, one study found that this increased absenteeism was due to co-morbidity rather than headache [83]. In addition to missed work, reduced productivity while at work with a headache accounts for an equal or greater work loss than actual missed days, equivalent to an additional 5 days or more per year (Tab. 3).

Conclusion

Migraine is a common condition that is associated with high rates of disability during attacks. As migraine is highly prevalent during the peak productive working years, economic costs to the individual, employer and society are comparable to, or greater than, other chronic conditions often thought of as being more disabling.

Table 3 - Annualized lost workdays and reduced productivity due to migraine

Ref.	Country	Lost work days	LWDE*	Total (LWD + LWDE)	Comment
[91]	UK	1.9	4.4	6.3	
Lipton (unpubl.)	US	3.1	16.6	18.8	Based on migraineurs with at least 6 attacks per year
[72]	Denmark				43% had at least 1 absent day
[92]	US	3.2	4.9	8.1	
[93]	US	3.8 (men) 8.3 (women)			"Severe" headache used as screening question; includes lost work of homemakers
[23]	UK	2.7	15.4	15.8	Based on migraineurs with at least 6 attacks per year
[32]	Netherlands	4.2	8.9	13.1	
[82]	US	2.8	5.6	8.4	Population-based 3-month diary study
[55]	UK	2.0	5.5	7.5	Hospital employees; 81% women
[90]	Spain	0.8 (men) 1.1 (women)		6.3 (men) 9.5 (women)	Employees of 11 large companies
[75]	UK				24% had at least 1 absent day
[78]	US	14.8	25.7	40.5	

**LWDE (lost workday equivalents) = equivalent number of days lost due to reduced productivity while at work*

References

1 Lipton RB, Stewart WF, Diamond S et al (2001) Prevalence and burden of migraine in the United States: data from the American Migraine Study II. *Headache* 41: 646–657

2 Murray CJ, Lopez AD (1997) Regional patterns of disability-free life expectancy and disability-adjusted life expectancy: global burden of disease study. *Lancet* 349: 1347–1352

3 Scher AI, Stewart WF, Lipton RB (1999) Migraine and headache: a meta-analytic approach. In: IK Crombie (ed): *The epidemiology of pain.* International Association for the Study of Pain Press, Seattle, 159–170

4 Stewart WF, Shechter A, Rasmussen BK (1994) Migraine prevalence. A review of population-based studies. *Neurology* 44: S17–S23
5 Headache Classification Committee of the International Headache Society (1988) Classification and diagnostic criteria for headache disorders, cranial neuralgias and facial pain. *Cephalalgia* 8 (Suppl 7): 1–96
6 Arregui A, Cabrera J, Leon-Velarde F et al (1991) High prevalence of migraine in a high-altitude population. *Neurology* 41: 1668–1669
7 Barea LM, Tannhauser M, Rotta NT (1996) An epidemiologic study of headache among children and adolescents of southern Brazil. *Cephalalgia* 16: 545–549
8 Cruz ME, Cruz I, Preux PM et al (1995) Headache and cysticercosis in Ecuador, South America. *Headache* 35: 93–97
9 Jaillard AS, Mazetti P, Kala E (1997) Prevalence of migraine and headache in a high-altitude town of Peru: a population-based study. *Headache* 37: 95–101
10 Lavados PM, Tenhamm E (1997) Epidemiology of migraine headache in Santiago, Chile: a prevalence study. *Cephalalgia* 17: 770–777
11 Alders EE, Hentzen A, Tan CT (1996) A community-based prevalence study on headache in Malaysia. *Headache* 36: 379–384
12 Roh JK, Kim JS, Ahn YO (1998) Epidemiologic and clinical characteristics of migraine and tension-type headache in Korea. *Headache* 38: 356–365
13 Sakai F, Igarashi H (1997) Prevalence of migraine in Japan: a nationwide survey. *Cephalalgia* 17: 15–22
14 Wang SJ, Liu HC, Fuh JL et al (1997) Prevalence of headaches in a Chinese elderly population in Kinmen: age and gender effect and cross-cultural comparisons. *Neurology* 49: 195–200
15 Wang SJ, Fuh JL, Young YH (2000) Prevalence of migraine in Taipei: a population-based survey. *Cephalalgia* 20: 566–572
16 Wong TW, Wong KS, Yu TS, Kay R (1995) Prevalence of migraine and other headaches in Hong Kong. *Neuroepidemiol* 14: 82–91
17 Abu-Arefeh I, Russell G (1994) Prevalence of headache and migraine in schoolchildren. *Br Med J* 309: 765–769
18 Raieli V, Raimondo D, Cammalleri R, Camarda R (1995) Migraine headaches in adolescents: a student population-based study in Montreal. *Cephalalgia* 15: 5–12
19 Stewart WF, Linet MS, Celentano DD et al (1991) Age- and sex-specific incidence rates of migraine with and without visual aura. *Am J Epidemiol* 134: 1111–1120
20 Bank J, Marton S (2000) Hungarian migraine epidemiology. Headache 40: 164–169
21 Bille B (1997) A 40-year follow-up of school children with migraine. *Cephalalgia* 17: 488–491
22 Stewart WF, Lipton RB, Liberman J (1996) Variation in migraine prevalence by race. *Neurology* 47: 52–59
23 Stewart WF, Lipton RB, Celentano DD, Reed ML (1992) Prevalence of migraine headache in the United States. Relation to age, income, race, and other sociodemographic factors. *J Am Med Assoc* 267: 64–69

24 Breslau N, Davis GC, Andreski P (1991) Migraine, psychiatric disorders, and suicide attempts: an epidemiologic study of young adults. *Psychiat Res* 37: 11–23

25 Stewart WF, Shechter A, Lipton RB (1994) Migraine heterogeneity. Disability, pain intensity, and attack frequency and duration. *Neurology* 44: S24–S39

26 Scher AI, Stewart WF, Liberman J, Lipton RB (1998) Prevalence of frequent headache in a population sample (abstract). *Headache* 38: 404

27 Silberstein SD, Lipton RB, Sliwinski M (1996) Classification of daily and near-daily headaches: field trial of revised IHS criteria. *Neurology* 47: 871–875

28 Saper JR (1990) Daily chronic headache. *Neurol Clinics* 8: 891–901

29 Mathew NT, Reuveni U, Perez F (1987) Transformed or evolutive migraine. *Headache* 27: 102–106

30 Scher AI, Stewart WF, Liberman J, Lipton RB (1998) Prevalence of frequent headache in a population sample. *Headache* 38: 497–506

31 Michel P, Pariente P, Duru G et al (1996) MIGACCESS: a population-based, nationwide, comparative survey of access to care in migraine in France. *Cephalalgia* 16: 50–55

32 Van Roijen L, Essink-Bot M, Koopmanschap MA et al (1995) Societal perspective on the burden of migraine in The Netherlands. *PharmacoEconomics* 7: 170–179

33 Merikangas KR (1991) Comorbidity of migraine and other conditions in the general-population of adults in the United States. *Cephalalgia* 11: 108–109

34 Essink-Bot ML, Van Royen L, Krabbe P et al (1995) The impact of migraine on health status. *Headache* 35: 200–206

35 Breslau N, Davis GC (1993) Migraine, physical health and psychiatric disorder: a prospective epidemiologic study in young adults. *J Psychiat Res* 27: 211–221

36 Andermann E, Andermann FA (1987) Migraine-epilepsy relationships: epidemiological and genetic aspects. In: FA Andermann, E Lugaresi (eds): *Migraine and epilepsy*. Butterworth-Heinemann, Boston, 281–291

37 Lipton RB, Ottman R, Ehrenberg BL, Hauser WA (1994) Comorbidity of migraine: the connection between migraine and epilepsy. *Neurology* 44: S28–S32

38 Ottman R, Lipton RB (1994) Comorbidity of migraine and epilepsy. *Neurology* 44: 2105–2110

39 Ottman R, Lipton RB (1996) Is the comorbidity of epilepsy and migraine due to a shared genetic susceptibility? *Neurology* 47: 918–924

40 Tzourio C, Iglesias S, Hubert JB et al (1993) Migraine and risk of ischaemic stroke: a case-control study. *Br Med J* 307: 89–292

41 Tzourio C, Tehindrazanarivelo A, Iglesias S et al (1995) Case-control study of migraine and risk of ischaemic stroke in young women. *Br Med J* 310: 830–833

42 Silberstein SD, Lipton RB, Breslau N (1995) Migraine: association with personality characteristics and psychopathology. *Cephalalgia* 15: 358–369

43 Merikangas KR, Angst J, Isler H (1990) Migraine and psychopathology. Results of the Zurich cohort study of young adults. *Arch Gen Psychiat* 47: 849–853

44 Stewart WF, Linet MS, Celentano DD (1989) Migraine headaches and panic attacks. *Psychosomatic Med* 51: 559–569

45 Stewart WF, Shechter A, Liberman J (1992) Physician consultation for headache pain and history of panic: results from a population-based study. *Am J Med* 92: S35–S40
46 Merikangas KR, Stevens DE, Angst J (1993) Headache and personality: results of a community sample of young adults. *J Psychiat Res* 27: 187–196
47 Breslau N (1992) Migraine, suicidal ideation, and suicide attempts. *Neurology* 42: 392–395
48 Breslau N, Davis GC (1992) Migraine, major depression and panic disorder: a prospective epidemiologic study of young adults. *Cephalalgia* 12: 85–90
49 Breslau N, Davis GC, Schultz LR, Peterson EL (1994) Migraine and major depression: a longitudinal study. *Headache* 34: 387–393
50 Magni G, Moreschi C, Rigatti-Luchini S, Merskey H (1994) Prospective study on the relationship between depressive symptoms and chronic musculoskeletal pain. *Pain* 56: 289–297
51 Stang P, Von Korff M, Galer BS (1998). Reduced labor force participation among primary care patients with headache. *J Gen Intern Med* 13: 296–302
52 Von Korff M, Le Resche L, Dworkin SF (1993) First onset of common pain symptoms: a prospective study of depression as a risk factor. *Pain* 55: 251–258
53 Breslau N, Schultz LR, Stewart WF et al (2000) Headache and major depression: is the association specific to migraine? *Neurology* 54: 308–313
54 Edmeads J, Findlay H, Tugwell P (1993) Impact of migraine and tension-type headache on life-style, consulting behaviour, and medication use: a Canadian population survey. *Can J Neurol Sci* 20: 131–137
55 Clarke CE, MacMillan L, Sondhi S, Wells NE (1996) Economic and social impact of migraine. *Q J Med* 89: 77–84
56 Kryst S, Scherl E (1994) A population-based survey of the social and personal impact of headache. *Headache* 34: 344–350
57 Schwartz BS, Stewart WF, Simon D, Lipton RB (1998) Epidemiology of tension-type headache. *J Am Med Assoc* 279: 381–383
58 Göbel H, Petersen-Braun M, Soyka D (1994) The epidemiology of headache in Germany: a nationwide survey of a representative sample on the basis of the headache classification of the International Headache Society. *Cephalalgia* 14: 97–106
59 Rasmussen BK (1992) Migraine and tension-type headache in a general population: psychosocial factors. *Int J Epidemiol* 21: 1138–1143
60 O'Brien B, Goeree R, Streiner D (1994) Prevalence of migraine headache in Canada: a population-based survey. *Int J Epidemiol* 23: 1020–1026
61 Pryse-Phillips W, Findlay H, Tugwell P et al (1992) A Canadian population survey on the clinical, epidemiologic and societal impact of migraine and tension-type headache. *Can J Neurol Sci* 19: 333–339
62 Osuntokun BO, Adeuja AO, Nottidge VA (1992) Prevalence of headache and migrainous headache in Nigerian Africans: a community-based study. *East Afr Med J* 69: 196–199

63 Tarlov AR, Ware JEJ, Greenfield S (1989) The medical outcomes study. An application of methods for monitoring the results of medical care. *J Am Med Assoc* 262: 925–930
64 Solomon GD, Skobieranda FG, Gragg LA (1993) Quality of life and well-being of headache patients: measurement by the medical outcomes study instrument. *Headache* 33: 351–358
65 Osterhaus JT, Townsend RJ, Gandek B, Ware JEJ (2000) Measuring the functional status and well-being of patients with migraine headache. *Headache* 34:337–343
66 Terwindt GM, Ferrari MD, Tijhuis M et al (1998) The impact of migraine on quality of life in the general population: the GEM study. *Neurology* 55: 624–629
67 Lipton RB, Liberman J, Kolodner K (2000) Migraine headache disability and quality-of-life: a population-based case-control study (abstract). *Headache* 39: 365
68 Lipton RB, Hamelsky SW, Kolodner K et al (2000) Migraine, quality of life, and depression: a population-based case-control study. *Neurology* 55: 629–635
69 Consumer Healthcare Products Association (2000). *OTC Sales Statistics 1995–1999*. AC Neilsen, www.ndmainfo.org
70 Hu XH, Markson LE, Lipton RB (1999) Burden of migraine in the United States – disability and economic costs. *Arch Intern Med* 159: 813–818
71 Schappert SM (1997) Ambulatory care visits to physician offices, hospital outpatient departments, and emergency departments: United States, 1995. *Vital health statistics*. National Center for Health Statistics, Hyattsville, Maryland
72 Rasmussen BK, Jensen R, Olesen J (1992) Impact of headache on sickness absence and utilisation of medical services: a Danish population study. *J Epidemiol Community Health* 46: 443–446
73 Schappert SM (1995) Office visits to neurologists: United States, 1991–92. *Vital health statistics*. National Center for Health Statistics, Hyattsville, Maryland
74 Forward SP, McGrath PJ, MacKinnon D et al (1998) Medication patterns of recurrent headache sufferers: a community study. *Cephalalgia* 18: 146–151
75 Mounstephen AH, Harrison RK (1995) A study of migraine and its effects in a working population. *Occupational Med* 45: 311–317
76 McCaig LF (2000) National hospital ambulatory medical care survey: 1998. *Vital health statistics*. National Center for Health Statistics, Hyattsville, Maryland
77 Celentano DD, Stewart WF, Lipton RB, Reed ML (1992) Medication use and disability among migraineurs: a national probability sample survey. *Headache* 32: 223–228
78 Osterhaus JT, Gutterman DL, Plachetka JR (1992). Healthcare resource and lost labor costs of migraine headache in the US. *PharmacoEconomics* 2: 67–76
79 Astin JA (1998) Why patients use alternative medicine: results of a national study. *J Am Med Assoc* 279: 1548–1553
80 Eisenberg DM, Davis RB, Ettner SL (1998) Trends in alternative medicine use in the United States, 1990–1997: results of a follow-up national survey. *J Am Med Assoc* 280: 1569–1575
81 Feeney A, North F, Head J et al (1998) Socioeconomic and sex differentials in reason for

sickness absence from the Whitehall II Study. *Occupational Environmental Med* 55: 91–98

82 Von Korff M, Stewart WF, Simon DJ, Lipton RB (1998) Migraine and reduced work performance: a population-based diary study. *Neurology* 50: 1741–1745

83 Michel P, Dartigues JF, Lindoulsi A, Henry P (1997) Loss of productivity and quality of life in migraine sufferers among French workers: results from the GAZEL cohort. *Headache* 37: 71–78

84 Clouse JC, Osterhaus JT (1994) Healthcare resource use and costs associated with migraine in a managed healthcare setting. *Ann Pharmacother* 28: 59–664

85 Goldfarb SD, Duncan BS, Dans PE, Sloan AS (1999) HMO direct costs and health care resources use after implementation of a monthly limit on sumatriptan. *Am J Health Syst Pharm* 56: 2206–2210

86 Lipton RB, Stewart WF, Von Korff M (1997) Burden of migraine: societal costs and therapeutic opportunities. *Neurology* 48: S4–S9

87 Streator SE, Shearer SW (1996) Pharmacoeconomic impact of injectable sumatriptan on migraine-associated healthcare costs. *Am J Managed Care* 2: 139–143

88 Lipton RB, Stewart WF, Simon D (1998) Medical consultation for migraine: results from the American migraine study. *Headache* 38: 87–96

89 Legg RF, Sclar DA, Nemec NL et al (1997) Cost-effectiveness of sumatriptan in a managed care population. *Am J Managed Care* 3: 117–122

90 Lainez J.M., Titus F., Cobaleda S., Leton E (1996) Socioeconomic impact of migraine (abstract). *Functional Neurol* 11: 133

91 Cull RE, Wells NEJ, Miocevich ML (1992) The economic cost of migraine. *Br J Med Economics* 2: 103–115

92 Schwartz BS, Stewart WF, Lipton RB (1997) Lost workdays and decreased work effectiveness associated with headache in the workplace. *J Occupational Environmental Med* 39: 320–327

93 Stewart WF, Lipton RB, Simon D (1996) Work-related disability: results from the American migraine study. *Cephalalgia* 16: 231–238

Presentation, pathogenesis, and treatment of migraine

Elizabeth W. Loder

Spaulding Rehabilitation Hospital, Pain Inpatient Unit, 125 Nashua Street, Boston, MA 02114-1198, USA

Presentation of migraine

Migraine is an intermittent headache syndrome whose pattern is generally easily recognized. It is comprised of a combination of signs and symptoms no one of which (other than headache) is absolutely required for diagnosis. Attacks can occur almost daily or very rarely, indicating that the threshold for experiencing an attack may vary from person to person. In an effort to standardize diagnosis, the International Headache Society has devised diagnostic criteria [1]. These have been useful for research purposes but perform less well in the clinical setting. Nonetheless, these criteria are widely accepted and used. They are summarized in Table 1.

Untreated or unsuccessfully treated, an attack of migraine is generally a long headache, lasting between 4 and 72 h. Migraine is often unilateral but it does not have to be. Confusion on this point is understandable, since the term "migraine" derives from the Greek term, "hemicrania," meaning one side of the head. However, many patients do not have unilateral headache and will describe the pain of a migraine attack as generalized, behind the eyes or on top of the head. It is also worth noting that the pain, if unilateral, can switch sides from attack to attack, or even within a single attack. This variability in the location of the pain is often said to be a reassuring sign that the underlying headache mechanism is benign, since a fixed structural abnormality leading to headache would more likely produce headache of unvarying location.

The pain of migraine is usually felt in the distribution of the trigeminal nerve and upper cervical roots and is often described as being over the temple(s). Pain-producing structures are innervated by the first division of the trigeminal nerve, connect with the trigeminocervical complex, and then to the dorsal horns of C1 and C2, finally transmitting signals to the brainstem, thalamus and cortex. Portions of the periaqueductal gray matter and locus coeruleus are hypothesized to serve as modulators of incoming pain signals and blood flow [2]. The placement of electrodes into the region of the periaqueductal gray has been noted to create a migraine-like headache, while in animals stimulation of the locus coeruleus reduced blood flow [3].

Migraine: A Neuroinflammatory Disease?, edited by Egilius L.H. Spierings and Margarita Sánchez del Río

Table 1 - Diagnostic criteria for migraine with and without aura (adapted from the International Headache Society [1])

Migraine without aura

A. At least five attacks fulfilling criteria B–D
B. Headache lasting 4 to 72 h (untreated or unsuccessfully treated)
C. Headache has at least two of the following characteristics:
   1. Unilateral location
   2. Pulsating quality
   3. Moderate or severe intensity (inhibits or prohibits daily activities)
   4. Aggravation by walking stairs or similar routine physical activity
D. During headache, at least one of the following occurs:
   1. Nausea and/or vomiting
   2. Photophobia and phonophobia
E. At least one of the following is present:
   1. History and physical and neurological examinations do not suggest an organic disorder
   2. History and/or physical and/or neurological examinations do suggest such disorder but it is ruled out by appropriate investigations
   3. Such disorder is present but migraine attacks do not occur for the first time in close temporal relation to the disorder

Migraine with aura

A. At least two attacks fulfilling criterion B
B. At least three of the following four characteristics are present:
   1. One or more fully reversible aura symptoms occur, indicating brain dysfunction
   2. At least one aura symptom develops gradually over more than 4 minutes or two or more symptoms occur in succession.
   3. No single aura symptom lasts more than 60 min
   4. Headache follows aura with a free interval of less than 60 mins (it also may begin before or simultaneously with the aura)
C. History, physical and neurological examination and, where appropriate, diagnostic tests exclude a secondary cause.

At its peak, the pain of migraine is often described as pounding or throbbing. Patients may use descriptive language, such as "a little man in my head with a hammer." The pain of a full-blown migraine attack is of moderate or severe intensity. In clinical trials of migraine medications, pain intensity has typically been rated on a 4-point scale of 0–3, with 0 being no pain, 1 mild pain, 2 moderate pain, and 3 severe pain. Pain ratings are inherently subjective; stoic patients will rarely report a pain level of 3, while more histrionic patients report all pain as very intense. Many clinicians find that it is more helpful to use reports of headache-induced disability as a proxy for pain intensity. Patients are asked to describe the impact of pain on their ability to carry out such daily activities as attending work or school, performing housework, or participating in leisure activities, and disability is rated as absent (0), mild (1), moderate (2), or severe (3). In most patients, the pain of migraine generally either interferes with daily activities (the patient might leave work early) or prohibits them altogether (the patient is unable to get out of bed to go to work). In addition, everyday activities that ordinarily do not cause pain, such as walking or climbing stairs, will aggravate the pain of a migraine headache.

In contrast to tension-type headache, in which pain is the principal feature of the headache disorder, migraine can be thought of as a "headache-plus" syndrome, in which characteristic accompanying features are the rule rather than the exception. Nausea is perhaps the most common symptom, sometimes preceding the headache or developing early, but more often paralleling the intensity of the pain. Vomiting is usually a late feature of an attack, but in some patients, it occurs early and can complicate management by making it difficult to take oral headache medications. Prolonged vomiting is particularly characteristic of childhood migraine and a childhood variant of migraine, known as "cyclic vomiting", is characterized by repeated bouts of abdominal pain and vomiting in which headache may be absent or only a minor feature. It often responds well to treatment with typical migraine regimens. Both nausea and vomiting usually become less prominent features of migraine as the patient ages.

Sensitivity to light and noise are also features of the migraine attack and may precipitate an attack or appear as a prodrome. Most migraine patients in the throes of a headache display a strong preference for quiet, dark rooms and prefer to lie motionless, avoiding even small movements, which can enhance the pain of the headache. Even interictally, migraine sufferers often report enhanced sensitivity to light, noise and smells. Indeed, in many ways migraine can be thought of as the result of a central nervous system that is exquisitely sensitive to many environmental stimuli, and has a lower threshold for activation than that of the average person.

Approximately 15% of patients with migraine experience migraine with aura. In this disorder, headache is preceded by transient focal neurological symptoms. Most common is some sort of transient visual disturbance, which varies from photopsia to complete visual loss. Paresthesias, numbness, weakness, confusion, vertigo and other symptoms can also occur, however. In the typical aura, the neurological event both begins and fades away gradually, with the entire aura episode lasting no more than one-half to one hour. As the aura fades, the headache begins. Most patients do not experience aura with every headache but the aura does tend to be relatively stereotyped when it occurs; thus, it is unusual for a patient to have a visual aura preceding one headache episode and hemiplegia preceding another. Some patients find the aura more troublesome than the headache that follows; this is particularly true when the aura comes on without warning and involves severe visual aberrations, which can interfere with driving or other activities.

Variations on the theme of typical aura occur and aura can in some circumstances be prolonged, lasting into or even beyond the headache. Similarly, as patients become older, it is not unusual to experience aura without a subsequent headache. A rare subtype of migraine with aura, familial hemiplegic migraine, has been linked in 50% of affected individuals to a locus on chromosome 19 and it is likely that the genetics of more common forms of migraine will soon be more completely understood.

Many patients experience premonitory events to migraine attacks that do not technically meet the criteria for aura, that is, they do not involve transient focal neurological symptoms. About 25% of patients will identify characteristic symptoms such as yawning, changes in mood or appetite prior to onset of a headache [4]. These may be related to hypothalamic involvement in the pathogenesis of a migraine attack and involve changes in dopaminergic transmission in the central nervous system. This is supported by the fact that polymorphisms have been identified in dopamine-receptor genes in migraineurs and that administration of a dopamine-receptor agonist, apomorphine, reproduces many of the typical migraine prodromal symptoms.

Pathogenesis of migraine

Disagreement exists about the exact cause of migraine. Current thinking is that underlying abnormalities in pain-control systems and enhanced central nervous system sensitivity to environmental triggers predispose patients to develop migraine. Genetic factors clearly play a role in this sensitivity. The familial occurrence of migraine has long been understood. Concordance for migraine in monozygotic twins is greater than that for dizygotic twins [5].

Various biochemical and neurovascular abnormalities may underlie the genetic susceptibility to migraine. Studies have suggested that variations within dopamine-receptor genes may affect migraine threshold [6]. In addition, studies of families with a rare autosomal dominant form of migraine with aura, known as familial hemiplegic migraine, have identified an amino acid substitution in the pore-forming and voltage-sensor regions of neuronal calcium channels. This has been traced to a missense mutation in a neuronal specific P/Q calcium channel gene and is present in roughly half of the families affected [7]. In total, four different missense mutations in the α_1 subunit have been identified. Abnormalities of genes encoded on chromosome 19 may also be involved in more common forms of migraine [8].

It has been postulated that the above mutations give rise to changes in channel expression that result in both a gain and a loss of neuronal function [9]. Neuronal calcium channels influence the release of serotonin in the midbrain and if these channels do not function appropriately, the release of serotonin and activation of the natural inhibitory system for migraine might be impaired [10]. Magnesium deficiency in the cortex of some migraineurs has also been noted and it is postulated that since magnesium interacts with calcium channels, a deficiency of magnesium may allow spreading depression to occur *via* activation of N-methyl-D-aspartate (NMDA) receptors, thus initiating a migraine aura or attack [11].

The biochemical and neurophysiological abnormalities noted above have in common a tendency to lead to marked central nervous system dysfunction in response to various environmental triggers, allowing initiation of a migraine attack. They may act, alone or in concert, to make the brain of the migraineur more susceptible to spreading depression. It is known, for example, that migraine patients have exaggerated responses to sensory stimuli, such as light, sound and smells. That this is so is supported by the study of occipital cortex responses to magnetic stimulation in migraineurs [12], as well as by studies of changes in visual processing [13] and auditory evoked potentials in migraineurs [14].

Several theories have been suggested to explain the cause of migraine. The first originated with Harold G. Wolff and is often referred to as the vascular theory of migraine [15]. This holds that vasoconstriction and ischemia account for the symptoms of the migraine aura, while reactive vasodilation activates primary sensory neurons. The belief in this hypothesis was reinforced by the fact that ergotamine, an effective anti-migraine medication, is also a potent vasoconstrictor. The blood vessels of the head are known to be richly innervated and to produce pain when dilated; in contrast, brain tissue itself is insensate. Vasodilation of blood vessels is, thus, theorized to activate perivascular sensory nerves and produce the pain of headache [16].

The neurogenic theory of migraine, in contrast, holds that the vascular changes that occur in migraine are a result of primary disturbances in the function of neurons. It also holds that neurogenic inflammation and plasma-protein extravasation generate headache pain through activation of primary sensory neurons [17]. This theory was bolstered by early blood-flow studies that appeared to contradict the vascular theory of migraine, demonstrating that although reduced blood flow did occur during the aura, it persisted well into the headache phase of a migraine [18]. Attempts to reconcile these two theories have led to the concept of migraine as a neurovascular phenomenon, in which the neural abnormalities that occur are fundamental and lead to reactive changes in vascular, sensory and pain-sensing pathways.

The slow evolution of the migraine aura has been noted for some time. Nearly 60 years ago, Lashley kept meticulous records of his own aura symptoms and speculated that the rate of progression of the cortical phenomenon producing it was around 3 mm per min [19]. The correlation between this calculated rate and that of spreading depression led Milner, a decade and a half later, to speculate that the phenomenon of spreading depression might account for the aura in migraine [20]. Spreading depression is a wave of cortical excitation, followed by depressed cortical activity, observed in animal studies.

Evidence to support or refute these competing theories has been difficult to come by, largely owing to the difficulties inherent in studying blood flow and neuronal function simultaneously during spontaneous, unprovoked migraine attacks. In addition, the migraine aura is a short and episodic event, which makes it difficult to schedule study of an attack, especially of its beginning phases. Recent advances in functional neuroimaging have made it possible to study both metabolic and hemodynamic changes during and between attacks of migraine. The techniques that have been used include Xenon blood flow, transcranial Doppler, positron emission tomography, single photon-emission computed tomography, magneto-encephalography, transcranial magnetic stimulation, and functional magnetic resonance imaging.

Xenon blood-flow studies have generally shown reductions in cerebral blood flow in posterior regions of the brain during aura-like symptoms induced by angiography. These changes persist for 30 min to 1 h and then normalize or remain focally decreased. A few investigators reported that the spread of oligemia occurred at a rate consistent with spreading depression, 2–3 mm/min. Although it is tempting to suppose that aura symptoms may be correlated with the amount and duration of blood-flow changes, it should be remembered that some patients with aura did not demonstrate a disturbance in cerebral blood flow in some series [21]. In addition, the point has been made that the blood-flow reductions observed (on the order of 17–35%) are well below the threshold required to cause ischemia (about 75%) [22].

Transcranial Doppler measures blood-flow velocity, but not blood volume, so it is not possible to know with certainty what the cerebral blood flow is. Results have been conflicting and difficult to interpret. Studies using single photon-emission computed tomography have generally shown areas of cerebral hypoperfusion, which seem to correlate with aura symptoms, while in migraine without aura, blood flow has generally appeared normal both between and during attacks. Positron emission tomography allows more precise measurement of cerebral blood flow. A spontaneous attack of migraine without aura has been captured in one subject: it showed bilateral and spreading blood-flow decreases beginning in the visual associative cortex, just a few minutes after onset of the headache, which gradually spread anteriorly [23]. Positron emission tomography has provided support for the theory of a migraine generator in the brain, responsible for the pain of migraine but not active during other types of head pain. Weiller et al. showed activation in the medial brainstem during attacks of migraine, which persisted despite pain relief from sumatriptan injection [24]. This area was not active when studied between headaches; in addition, head pain induced by injections of capsaicin in normal volunteers did not activate the brainstem region putatively involved in migraine [25].

Cutrer et al. have recently summarized much of the data from neuroimaging and blood-flow studies [26]. They concluded that the bulk of the evidence supports the idea that the visual aura is associated with "transient decreases in regional cerebral blood flow, which are most often posterior in origin" and that these changes migrate anteriorly over time. They comment that the magnitude of the blood-flow decline is much smaller than necessary to produce ischemic injury and that there is no evidence that migraine without aura is associated with significant blood-flow changes. They conclude that the evidence to date increasingly suggests fundamental differences between the "physiology of cortical neurons" and neuronal energy metabolism in patients who have migraine with aura, compared with those who have migraine without aura and between both of these groups and normal controls.

Trigeminovascular activation has been theorized to lead to release of various neuroinflammatory peptides, such as calcitonin gene-related peptide and substance P, which promote leakage of plasma proteins from blood vessels into surrounding tissue, referred to as plasma-protein extravasation. Experimentally, the release of these neuropeptides can be demonstrated with electrical stimulation of the trigeminal ganglion in animals [27] and is blocked by many compounds that are also effective antimigraine medications, such as ergots, triptans and indomethacin [28]. Levels of calcitonin gene-related peptide are elevated in the cranial circulation during attacks of migraine and normalize with successful treatment of migraine. Further evidence of the role of calcitonin gene-related peptide in migraine comes from the fact that infusion of the peptide can, in certain situations, precipitate migraine [29].

Finally, much interest has centered on the role of serotonin in the genesis of migraine. It has been known since the 1960s that circulating serotonin levels fall during a migraine attack and that infusions of serotonin could successfully abort an attack [30]. Serotonin receptors are found throughout the cranial circulation and to a lesser extent on vessels elsewhere. In addition, serotonin-1B and -D receptors have been found in the trigeminal nucleus caudalis of the cat in a distribution, which correlates with locations where proteins are seen after pain-produced activation of neurons in the sagittal sinus [31]. Recently, interest has centered on nitric oxide as the "final common pathway" through which the vasodilation of migraine occurs [32]. It is far from certain that only serotonergic pathways can activate this mechanism and it is likely that in the future, we will see increased interest in other pathways that may be involved in the pathogenesis of migraine.

Abortive migraine treatment

Most patients whose headaches are severe enough to seek medical attention and who are diagnosed with migraine will require treatment of some sort. The United States Headache Consortium has recently released evidence-based guidelines for non-pharmacological as well as abortive and preventive pharmacological treatment of migraine [33]. The guidelines are a useful summary of the quality of existing evidence for many therapies and can be accessed through the web site of the American Academy of Neurology (www.aan.com). Unfortunately, evidence of any kind is lacking for many of the treatment decisions that must be made in a clinical setting and the guidelines are, thus, only partially useful. In addition, the criteria for inclusion of studies in the evidence review favored the newer medications that have been more rigorously studied; the exclusion of many older medications is a reflection of the generally lower scientific quality of the clinical studies done at the time they were introduced and not of their intrinsic value.

Non-pharmacological approaches

For many patients, lifestyle modification and non-pharmacological treatment strategies can be a helpful starting point, although rarely adequate treatment by themselves. Identification and elimination of typical headache triggers can be helpful in minimizing attacks that occur, although it should be recognized that not all triggers are avoidable. Many triggers are probably weak provocative agents and need to occur in concert to cause an attack. In addition, scientific evidence is lacking for many commonly mentioned headache triggers. Over-attention to headache triggers with resultant somatic preoccupation should not be encouraged. Table 2 lists commonly mentioned migraine triggers.

When the patient averages two or fewer headaches per week, abortive treatment alone will often prove successful. For patients whose headaches do not respond optimally to abortive treatment or who have more than two headache episodes per week, preventive treatment aimed at decreasing attack frequency may be necessary. The majority of patients have a preference for abortive treatment alone and where possible, this should be the first strategy attempted. The introduction of highly effective, disease-specific, abortive treatments underscores the importance of this statement. Currently available preventive treatments for migraine have side-effect profiles that can be difficult for young, otherwise healthy people to tolerate and many patients have a philosophical preference to avoid taking medication when they are not experiencing an attack. Finally, many migraine sufferers are women of childbearing age; the high rate of unintended pregnancies (around 50% in the United States) makes avoidance of daily medication where possible important in this group.

Some general principles of abortive treatment should be borne in mind. In general, early treatment at a mild stage of the attack is more successful than treatment employed late in an attack. Such early intervention has been shown to increase the number of patients who are pain-free, as opposed to being simply improved, at both 2 and 4 h after treatment. It has also been shown to reduce the requirement for a second dose of medication, decrease functional disability, and improve other desirable outcome measures [34]. This is so probably because the trigeminovascular sys-

Table 2 - Frequently mentioned migraine triggers

Emotional stress
Prolonged physical exertion
Relaxation after stress ("let-down headache")
Lack of sleep; oversleeping
Skipping meals
Alcohol, especially red wine and beer
Preserved meats
Aged cheeses
Orgasm
Vasodilating agents, such as nitrates, sildenafil

tem is not fully "wound up" yet and, therefore, easier to shut down. Thus, intervention at a mild stage of headache will prove cost effective for the bulk of migraine patients and should be encouraged. Adequate doses of medication should be used and associated symptoms, such as nausea, treated separately when necessary. In general, abortive medications for migraine should not be used more than 2–3 days per week, to avoid the possibility of rebound or analgesic-induced headache. Since migraine is characterized by recurrent episodes of headache over a lifetime in persons who are otherwise generally healthy, the aims of treatment go beyond simply providing pain relief and include avoiding treatment-related disability.

Selection of appropriate medication

Abortive treatment for migraine can be broadly divided into non-specific and specific. Sedative and analgesic medications that cover up the symptoms of an attack fall into the first category, while ergots and the new category of triptans fall into the latter. As their use is all too frequently associated with addiction, dependence, or sedation, it is preferable to avoid medications that contain short-acting barbiturates, benzodiazepines, or opioids for the abortive treatment of migraine. It is generally not possible to predict accurately those patients at risk for the development of these serious overuse syndromes; once established, these behavioral patterns are notoriously difficult to change. The dangers associated with these non-specific treatments and the availability of newer, more specific treatments that do not confer such risks, make the use of migraine-specific medications, such as the triptans, increasingly the standard of care for abortive migraine treatment.

Ergotamine has been available since the early decades of the 1900s for the treatment of migraine. It has complex receptor pharmacology, with prolonged vasoconstriction and nausea being especially undesirable effects of treatment. The triptan class of medications has proved an advance over the ergots, with response rates of around 70–80%, depending upon the agent used, dose and route of administration. Although it is the current standard of care for abortive migraine treatment, this class of medications is not perfect. Headache recurrence, lack of response in up to 30% of patients, and chest symptoms are among their limitations. Both the ergots and triptans are theorized to work *via* agonist activity at serotonin-1B and -1D receptors. Activation of these receptors is postulated to constrict blood vessels, decrease the activation of trigeminal neurons, and prevent central sensitization. The small degree of coronary vasoconstriction caused by the triptans at pharmacologically relevant doses has no clinical consequences in healthy patients. However, it accounts for the fact that these medications are contraindicated in patients with known coronary artery disease or risk factors for coronary artery disease. Over 10 years of clinical experience with this class of medications, however, has proved reassuring regarding their safety and tolerability. The largest study to date of triptan safety approximated "real world" conditions and provided reassuring evidence for the safety of these medications [35].

The choice of abortive treatment depends on many factors, including the intensity of the headache attack and any associated features (Tab. 3). Nausea or vomiting may make use of oral medications difficult or the presence of co-existing disorders may preclude use of various categories of medication. Simple analgesics available without prescription, with or without caffeine, may be appropriate treatment for mild or moderate migraine headaches but more severe attacks will usually require prescription medications.

Opioid medications should generally be reserved for patients who are not candidates for more specific medications or who have failed with the use of other medications. Short-acting opioids are most appropriately used for the majority of migraine sufferers as "rescue" therapy for individual attacks that do not respond to treatment. In general, the use of meperidine should be avoided, as it is neuro-exciting and may predispose to seizures. If opioids are used, adequate doses should be employed and adjunctive treatment of nausea considered.

Stage-specific treatment

To conceptualize abortive treatment strategies, it can be useful to arbitrarily divide a migraine attack into the stages of prodrome, aura, headache, and postdrome. The prodrome consists of subtle signs or symptoms that, if recognized, can alert the patient and physician to an impending attack. Many patients report such things as changes in appetite, mood, bowel habits, or mild muscle pain. Early intervention during prodrome may hold promise as a means of aborting a full-blown attack. Clinical experience indicates that the prodrome may often be successfully treated

Table 3 - Abortive medications for migraine

Non-prescription medications for mild or moderate headache

- Aspirin
- Acetaminophen
- Ketoprofen
- Naproxen sodium
- Aspirin-acetaminophen-caffeine

Prescription medications for mild or moderate headaches

- Midrin
- Tramadol
- Aspirin (or acetaminophen)-butalbital-caffeine combinations

Prescription medications for moderate or severe headaches

- Sumatriptan
- Rizatriptan
- Zolmitriptan
- Naratriptan
- Ergotamine
- Opioids

Medications to treat associated nausea or vomiting

- Metoclopramide
- Chlorpromazine
- Hydroxyzine
- Prochlorperazine

with the use of modest doses of milder medications than are necessary once an attack is fully underway. Anti-inflammatory analgesics seem to be particularly useful in this setting. Few interventions have been shown to be successful in aborting or preventing aura. In contrast to smaller doses of milder medications that may be successful in treating prodrome, attack-based therapy for established headaches may require larger doses of migraine-specific medications. Treatment of the postdrome is often ignored, yet anecdotes suggest that anti-inflammatory analgesics may play a role here as well.

Preventive migraine treatment

Patients whose headaches do not respond well to abortive treatment or who have very frequent headaches may be candidates for preventive treatment of migraine. Opinion differs on the number of headaches beyond which preventive treatment should be considered. In the past, some authorities have advocated preventive treatment for any patient with more than two headaches per month, while more recently others have suggested that it need not be considered necessary until the patient averages two headaches per week. This change in threshold for the institution of preventive treatment is likely to reflect recognition of the improved efficacy and reliability of the newer abortive medications, which can in some cases obviate the need for preventive treatment. Certainly, given the limited effectiveness and side-effect profiles of many of the preventive medications, patient preference is usually to avoid their use unless necessary.

Preventive treatment is given daily, with the intention of reducing the frequency and intensity of migraine episodes, as well as improving response to abortive medications. A number of classes of medications have proved useful for this purpose but it should be remembered that the bar for demonstration of effect is set relatively low. In general, an agent is considered an effective migraine-preventive medication if it decreases headache frequency by at least 50% in 50% of patients.

The choice of preventive medications is often made based on side-effect profile and patient co-morbidity. For example, migraine patients with hypertension or essential tremor may find benefit for both disorders with the use of beta-blockers. In contrast, in patients who have pre-existing asthma or depression, this class of medications should be avoided. Patients with co-morbid bipolar disorder or epilepsy might be well managed with divalproex.

The use of long-acting, scheduled opioids for chronic treatment of headache is controversial. The best evidence of their usefulness to date comes from a long-term study in Michigan. This 5-year study suggests that only a minority of headache patients will benefit long-term from this form of treatment and that the need for concomitant treatment of headache does not decrease. Thus, this treatment, while increasingly accepted for other kinds of chronic, non-malignant pain syndromes, appears to have only a limited role in the treatment of chronic headache disorders. Table 4 lists commonly used preventive medications and their doses for migraine.

Special treatment situations

Special considerations may apply in the treatment of certain sub-populations of migraine patients. These include women who are attempting pregnancy or who are pregnant, headaches that are predictably associated with the menstrual cycle, headaches occurring during the climacteric, and headaches occurring in children or the elderly.

Table 4 - Preventive medications for migraine

Medication	Typical dose
Tricyclics (low dose), for example, amitriptyline	25–150 mg per day
Beta-blockers, for example, propranolol*	80–160 mg per day
Divalproex sodium*	250–500 mg twice daily
Calcium-channel blockers, for example, verapamil	80–240 mg per day
Methysergide*	4–8 mg per day
Cyproheptadine	4–8 mg twice daily
Non-steroidal anti-inflammatory analgesics, for example, naproxen sodium	550 mg twice daily

**Approved by the Food and Drug Administration for preventive treatment of migraine*

Pregnancy

For the majority of women with migraine, some improvement in both headache frequency and intensity can be expected during the second and third trimesters of pregnancy. Because organogenesis is occurring during the first trimester, most women and their physicians are highly motivated to avoid medication treatment of migraine during this time. However, if headaches have not improved by the beginning of the second trimester, they are unlikely to and treatment may be necessary. Acetaminophen and opioids are safe choices when abortive treatment cannot be avoided and most experts would recommend the use of beta-blockers or low-dose tricyclics, if preventive treatment is needed [36]. The triptans are not currently recommended for use in pregnancy but evidence to date from an ongoing prospective pregnancy registry maintained by the manufacturer (GlaxoSmithKline, Internal report, 2001), as well as evidence from other sources, is reassuring that the medication may ultimately prove safe for use in pregnant patients [37]. The importance of avoidance of trigger factors and non-pharmacological treatment during pregnancy cannot be emphasized enough [38].

Menstrually associated migraine

A substantial minority of women with migraine notes a link between menstrual periods and headache. Currently, abortive treatment of menstrual migraine for most patients with the disorder is similar to that of non-menstrual attacks. However, non-steroidal anti-inflammatory analgesics may be more useful, in view of the role that prostaglandins play in other menstrually related symptoms. Ergots and triptans as well as non-specific medications, such as butalbital compounds and opioids, are employed successfully in menstrual migraine. Contrary to widely held belief, there is substantial evidence that most of these treatments work equally well in migraine associated with menstruation, as they do in non-menstrual migraine. Rizatriptan [39], zolmitriptan [40], sumatriptan [41] and eletriptan [42] have all been shown to be equally effective in menstrually and non-menstrually associated migraine and similar evidence exists for the non-specific combination analgesics, containing aspirin, caffeine and acetaminophen [43].

Short- or long-term preventive treatment with non-steroidal anti-inflammatory analgesics, ergots, selective serotonin re-uptake inhibitors, calcium-channel blockers, or magnesium supplements has been advocated for menstrual migraine as well. Naproxen sodium in a dose of 550 mg twice daily has been best studied [44, 45]. Some authorities advocate perimenstrual increases in the dose of a preventive medication being used throughout the month, while others advocate the use of short-term perimenstrual prevention only around the menstrual period. The latter is often recommended for women whose menstrually related headaches alone are resistant to therapy.

Since estrogen withdrawal has been implicated as the cause of menstrually associated migraine, there has long been enthusiasm for the use of estrogen supplementation to prevent or blunt the premenstrual fall in estrogen levels. Results of clinical studies have been variable, however, and there may be a critical level of estrogen necessary for efficacy. This was suggested by a study by Pradalier in which 25-µg patches were not effective in preventing menstrual migraine, while 100-µg patches were [46]. Still other studies found that the 50-µg patch was not effective [47].

Many preventive hormonal treatments for refractory menstrual migraine are based on expert experience or small, open-label studies. Some of those suggested are the use of bromocriptine [48], tamoxifen [49], danazol [50], and gonadotrophin-releasing hormone analogues with add-back estrogen therapy [51]. It is important to remember that the risk-to-benefit ratio of most of these treatments has not been carefully studied.

Menopause

Some women whose headaches previously followed a stable pattern will find that the hormonal changes of the climacteric make headaches less predictable or more intense. They may require preventive treatment for a period of time, until the pattern stabilizes. Questions regarding the advisability of hormone replacement therapy in women with a history of migraine cannot be answered on the basis of rigorous studies. However, clinical experience suggests that most women with migraine will not experience aggravation of their headaches with the use of daily low-dose estrogen hormonal replacement. The older, interrupted estrogen regimens, in which treatment was discontinued for 5–7 days at the end of the month, were more likely to produce estrogen-withdrawal headache. They may account for the commonly held view that patients with migraine should not use hormone replacement. Some women may find the stable hormonal levels of estrogen replacement beneficial for migraine.

Elderly

The majority of patients with migraine note an improvement in both headache frequency and intensity over time, so that headache is often less troublesome when they are older. However, there are exceptions to this rule and treatment can be complicated. Older patients have cardiac risk factors, which may make the use of triptans or other highly effective treatment problematic. They are also at higher risk for structural or other medical causes of headache, such as temporal arteritis, and care must be taken not to overlook these diagnostic possibilities in someone with a prior history of migraine.

Children

Migraine is a condition that frequently begins in childhood or adolescence. In children, headache attacks may be shorter than in adults and the period of disability and illness may be significantly compressed. This makes early treatment all the more important. Studies are ongoing to show effectiveness of new treatments, such as triptans, in children but the Food and Drug Administration does currently not approve these medications for use in children and adolescents; still, their off-label use in this population is common. Many physicians prefer specific medications, such as the triptans, to non-specific medications, such as barbiturates, with their high potential for abuse and sedation. Children often respond well to avoidance of trigger factors and interventions, such as biofeedback. Wherever possible, treatment should be incorporated in the child's routine and absences from school or extracurricular activities minimized.

References

1 International Headache Society (1988) Classification and diagnostic criteria for headache disorders, cranial neuralgias and facial pain. *Cephalalgia* 8 (Suppl 7): 1–96

2 Goadsby PJ (2000) Headache. In: DA Warrell, TM Cox, JD Firth (eds): *Oxford Textbook of Medicine*. Oxford Press, Oxford, 343–349

3 Goadsby PJ, Lambert GA, Lance JW (1982) Differential effects on the internal and external carotid circulation of the monkey evoked by locus coeruleus stimulation. *Brain Res* 249: 247–254

4 Drummond PD, Lance JW (1984) Neurovascular disturbances in headache patients. *Clin Exp Neurol* 20: 93–99

5 Honkasalo ML, Kaprio J, Winter T et al (1995) Migraine and concomitant symptoms among 8,167 adult twin pairs. *Headache* 35: 70–78

6 Peroutka SJ, Wilhoit T, Jones K (1997) Clinical susceptibility to migraine with aura is modified by dopamine D2 receptor (DR_{D2}) Nco1 alleles. *Neurology* 49: 201–206

7 Ophoff RA, Terwindt GM, Vergouwe MN et al (1996) Familial hemiplegic migraine and episodic ataxia type-2 are caused by mutations in the Ca^{2+} channel gene CACNL1A4. *Cell* 87: 543–552

8 Nyholt DR, Lea RA, Goadsby PJ et al (1998) Familial typical migraine: linkage to chromosome 19p13 and evidence for genetic heterogeneity. *Neurology* 50: 1428–1432

9 Hans M, Luvisetto S, Williams ME et al (1999) Functional consequences of mutations in the human alpha 1A calcium channel subunit linked to familial hemiplegic migraine. *J Neurosci* 19: 1610–1619

10 Yakhinista VA, Pilyavskii AI, Limansky YP, Bulgakova NV (1996) Modulation of the activity of midbrain central gray substance neurons by calcium channel antagonists *in vitro*. *Neuroscience* 70:159–167

11 Ramadan NM, Halvorson H, Vande-Linde A et al (1989) Low brain magnesium in migraine. *Headache* 29: 416–419

12 Aurora SK, Cao Y, Bowyer SM, Welch KMA (1999) The occipital cortex is hyperexcitable in migraine experimental evidence. *Headache* 39: 469–476

13 Wray SH, Mijovic-Prelec D, Kosslyn SM (1995) Visual processing in migraineurs. *Brain* 118: 25–35

14 Wang W, Schoenen J (1998) Interictal potentiation of passive "oddball" auditory event-related potentials in migraine. *Cephalalgia* 18: 261–265

15 Wolff HG (1963) Headache and other head pain. Oxford University Press, New York

16 Moskowitz MA (1991) The visceral organ brain: implications for the pathophysiology of vascular head pain. *Neurology* 41: 182–186

17 Moskowitz MA, MacFarlane R (1993) Neurovascular and molecular mechanisms in migraine headaches. *Cerebrovasc Brain Metab Rev* 5: 159–177

18 Cutrer FM, Sorensen AG, Weisskoff RM et al (1998) Perfusion-weighted imaging defects during spontaneous migraine aura. *Ann Neurol* 43: 25–31

19 Lashley KS (1941) Patterns of cerebral integration indicated by the scotomas of migraine. *Arch Neurol Psychiat* 46: 331–339

20 Milner PM (1958) Note on a possible correspondence between the scotomas of migraine and spreading depression of Leão. *Electroencephalogr Clin Neurophysiol* 10: 705

21 Andersen AR, Friberg L, Olsen TS, Olesen J (1988) SPECT demonstration of delayed hyperemia following hypoperfusion in classic migraine. *Arch Neurol* 45: 154–159

22 Lauritzen M, Olesen J (1984) Regional cerebral blood flow during migraine attacks by xenon-133 inhalation and emission tomography. *Brain* 107: 447–461

23 Woods RP, Iacoboni M, Mazziotta JC (1994) Bilateral spreading cerebral hypoperfusion during spontaneous migraine headache. *N Engl J Med* 331: 1689–1692

24 Weiller C, May A, Limmroth V et al (1995) Brainstem activation in spontaneous human migraine attacks. *Nature Med* 7: 658–660

25 May A, Kaube H, Buchel C et al (1998) Experimental cranial pain elicited by capsaicin: a PET study. *Pain* 74: 61–66

26 Cutrer FM, O'Donnell AO, Sanchez del Rio M (2000) Functional neuroimaging: enhanced understanding of migraine pathophysiology. *Neurology* 55 (Suppl 2): S36–S45

27 Markowitz S, Saito K, Moskowitz MA (1987) Neurogenically mediated leakage of plasma proteins occurs from blood vessels in dura mater but not brain. *J Neurosci* 7: 4129–4136

28 Buzzi MG, Moskowitz MA (1991) Evidence for 5-$HT_{1B/1D}$ receptors mediating the antimigraine effect of sumatriptan and dihydroergotamine. *Cephalalgia* 11: 165–168

29 Lassen L, Jacobsen V, Petersen P et al (1998) Human calcitonin gene-related peptide (hCGRP)-induced headache in migraineurs (abstract). *Eur J Pharmacol* 5 (Suppl 3): S63

30 Goadsby PJ, Edvinsson L (1994) Human *in vivo* evidence for trigeminovascular activation in cluster headache: neuropeptide changes and effects of acute attack therapies. *Brain* 17: 424–434

31 Kaube H, Keay KA, Hoskin KL et al (1993) Expression of c-fos-like immunoreactivity in the caudal medulla and upper cervical spinal cord following stimulation of the superior sagittal sinus in the cat. *Brain Res* 629: 95–102

32 Olesen J, Thomsen LL, Iversen H (1994) Nitric oxide is a key molecule in migraine and other vascular headaches. *Trends Pharmacol Sci* 15: 149–153

33 U.S. Headache Consortium (2000) American Academy of Neurology Headache Guidelines at: http://www.aan.com/public/practiceguidelines/headache_gl.htm

34 Cady RK, Sheftell F, Lipton RB et al (2000) Effect of early intervention with sumatriptan on migraine pain: retrospective analyses of data from three clinical trials. *Clin Ther* 22: 1035–1048

35 O'Quinn S, Davis RL, Gutterman D et al (1999) Prospective large-scale study of the tolerability of subcutaneous sumatriptan injection for acute treatment of migraine. *Cephalalgia* 19: 223–231

36 Wood A (2000) Drugs in pregnancy. *New Engl J Med* 338: 1130–1137

37 Shuhaiber S, Pastuszak A, Schick B et al (1998) Pregnancy outcome following first trimester exposure to sumatriptan. *Neurology* 51: 581–583

38 Scharff L, Marcus DA, Tuck DC (1996) Maintenance of effects in the nonmedical treatment of headaches during pregnancy. *Headache* 36: 285–290

39 Silberstein SD, Massiou H, LeJeunne C et al (2000) Rizatriptan in the treatment of menstrual migraine. *Obstetrics Gynecol* 96: 237–242

40 Loder E, Silberstein S (1998) Clinical efficacy of 2.5 and 5 mg zolmitriptan in migraine

associated with menses or in patients using non-progestogen oral contraceptives (abstract). *Neurology* 50 (Suppl 4): A341

41 Salonen R, Saiers J (1999) Sumatriptan is effective in the treatment of menstrual migraine: a review of prospective studies and retrospective analyses. *Cephalalgia* 19: 16–19.

42 Massiou H, Pitei D, Poole PH, Sikes C (2000) Efficacy of eletriptan for the treatment of migraine in women with menstrually associated migraine, and in women on contraceptives or hormone replacement therapy: meta-analyses of randomized clinical trials (abstract). *Cephalalgia* 20: 435

43 Silberstein SD, Armellino JJ, Hoffman HD et al (1999) Treatment of menstruation-associated migraine with the nonprescription combination of acetaminophen, aspirin, and caffeine: results from three randomized, placebo-controlled studies. *Clin Ther* 21: 475–491

44 Szekely B, Merryman S, Croft H et al (1989) Prophylactic effects of naproxen sodium on perimenstrual headache: a double blind placebo controlled study. *Cephalalgia* 9 (Suppl 10): 452–453

45 Sances G, Martignoni E, Fioroni L et al (1990) Naproxen sodium in menstrual migraine prophylaxis: a double-blind, placebo controlled study. *Headache* 30: 705–709

46 Pradalier A, Vincent D, Beaulieu PH et al (1994) Correlation between oestradiol plasma level and therapeutic effect on menstrual migraine. In: FC Rose (ed): *New advances in headache research*. Smith-Gordon, London, 129–132

47 Pfaffenrath V (1993) Efficacy and safety of percutaneous estradiol vs. placebo in menstrual migraine (abstract). *Cephalalgia* 13: 168

48 Herzog AG (1997) Continuous bromocriptine therapy in menstrual migraine. *Neurology* 48: 101–102

49 O'Dea PK, Davis EH (1990) Tamoxifen in the treatment of menstrual migraine. *Neurology* 40: 1470–1471

50 Powles TJ (1986) Prevention of migraineous headaches by tamoxifen. *Lancet* 2: 1344.

51 Silberstein SD, Merriam GR (1993) Sex hormones and headache. *J Pain Symptom Management* 8: 98–114

Neurogenic inflammation in the pathogenesis of migraine

David W. Dodick

Mayo Clinic, Department of Neurology, 13400 East Shea Boulevard, Scottsdale, AZ 85259, USA

Introduction

Migraine is a prevalent and disabling disorder, which affects approximately 12% of the population of the United States [1]. The pathogenesis of the headache involves pain-sensitive cranial structures, possibly including the dura mater and large cerebral blood vessels and their trigeminal and cervical innervation [2]. The innervation of these structures is mediated primarily by the tentorial nerve, a branch of the ophthalmic nerve or first division of the trigeminal nerve, which also provides cutaneous innervation for the forehead and anterior vertex.

Vasodilation as the cause of the pain of migraine headache has been considered since the early work by Harold G. Wolff and colleagues [3]. However, over the past two decades, under the influence of the work by Michael A. Moskowitz [4–8], focus has shifted from the blood vessels to their sensory innervation. In particular, the nociceptive fibers of the trigeminal nerve have been extensively studied, since most nociceptive information from pain-sensitive cranial structures traverses these fibers to access the central nervous system. In general, the study of the nociceptive trigeminal fibers has been based on two fundamental functions of these fibers:

1. The transmission of sensory (pain) information to the central relay nucleus of the trigeminal and cervical innervation of the head, the trigeminal nucleus caudalis; and
2. The capacity, like all sensory nerve fibers, to promote inflammation within the innervated tissues by releasing peptides from perivascular nerve endings, causing vasodilation and an increase in vascular permeability.

Anatomic tracing and immunohistochemical studies, in animals and man, have provided evidence for the existence of perivascular sensory nerve fibers. They have also provided evidence for the presence of vesicles within the fibers, containing substance P and calcitonin gene-related polypeptide, as well as for the presence of messenger RNA encoding for these neuropeptides in the cell bodies of the trigeminal ganglion

Migraine: A Neuroinflammatory Disease?, edited by Egilius L.H. Spierings and Margarita Sánchez del Río

[9, 10]. The neuropeptides, manufactured in the cell bodies, are transported down the nerve fibers to a plexus of small, unmyelinated nerve fibers, characteristic of C-fibers, in the adventitial layer of the blood vessels.

Neurogenic inflammation

Moskowitz and colleagues have shown that, in the rat and guinea pig, electrical stimulation of the trigeminal ganglion results in inflammation of the dura mater and extracranial tissues but not of the brain. The inflammation is neurogenic in origin because it is caused by the (antidromic) activation of the sensory nerve fibers [6] and mediated by the release of neuropeptides, in particular calcitonin gene-related peptide and substance P, from the nociceptive Aδ- and C-fibers, respectively [7]. Although both peptides are vasoactive, calcitonin gene-related peptide is the most potent vasodilator of the two, acting directly through activation of specific receptors on the vascular smooth-muscle cell. On the other hand, substance P, and not calcitonin gene-related peptide, mediates plasma-protein extravasation (leakage of albumin) from blood vessels, most likely at the level of the post-capillary venule. The peptides released from the nociceptive nerve fibers initiate a cascade of events, which leads to vasodilation and leakage of plasma proteins through the formation of endothelial microvilli, vesicles, and vacuoles within the post-capillary venules, with subsequent degranulation of mast cells and aggregation of blood platelets.

Neurogenic inflammation in man

Substance P

The suggestion that neurogenic inflammation is involved in the pathogenesis of the migraine headache was strengthened by studies demonstrating that medications beneficial in the treatment of migraine effectively block experimentally induced plasma-protein extravasation. These medications include the ergots, methysergide, triptans, such as sumatriptan, non-steroidal anti-inflammatory analgesics, such as aspirin and indomethacin, gamma-aminobutyric acid (GABA)-receptor agonists, such as divalproex, and CP-122288, a highly potent analogue of sumatriptan [11].

Although neurogenic inflammation has not been definitively demonstrated in man during migraine headache, evidence for involvement of this mechanism comes from a number of observations. First, levels of both substance P and calcitonin gene-related peptide are markedly increased in blood drawn from the *internal* jugular vein after thermo-coagulation of the trigeminal ganglion [12, 13]. Second, not only are the serotonin-1B/D-receptor agonists or triptans effective inhibitors of neurogenic inflammation in animals, their efficacy in migraine and cluster headache also coincides with their ability to normalize elevated levels of calcitonin gene-related peptide [14].

Substance P is one of three endogenous substances, called tachykinins, the others being neurokinin A and B. It is distributed throughout the cranial vasculature and in the trigeminal sensory nerve fibers. Of the three tachykinin-receptors that have been identified (NK-1, NK-2, NK-3), substance P has the highest affinity for the neurokinin-1 receptor. Since substance P is thought to be the key mediator of plasma-protein extravasation, compounds that block or antagonize the effects of this peptide were developed for the treatment of migraine.

Studies in animals with high-affinity, non-peptide neurokinin-1-receptor antagonists (CP-99994, GR-205171, LY-303870, RP-67580 and RPR-100893) demonstrated powerful inhibition of plasma-protein extravasation in the dura mater [15–18]. They also showed powerful inhibition of expression of messenger RNA for c-fos as well as fos protein in the trigeminal nucleus caudalis, following electrical stimulation of the trigeminal ganglion or noxious chemical stimulation of the dura mater. These observations in animals suggested that neurokinin-1-receptor antagonists might have antimigraine effects peripherally, through blocking plasma-protein extravasation, and centrally, by inhibiting nociceptive neurons.

Unfortunately, initial results from placebo-controlled trials were disappointing, showing no difference between LY-303870 and placebo in providing pain relief in migraine headache at 2 h after dosing [19]. As absorption of this particular neurokinin-1-receptor antagonist is decreased during migraine headache to less than 10% of that in fasting non-headache controls, it was felt that the plasma concentrations achieved in this study might not have been adequate to confer efficacy. However, recent placebo-controlled trials have demonstrated that neither GR-205171 nor LY-303870 administered intravenously was effective in reducing the intensity of migraine headache [20]. Since the compounds were administered intravenously, lack of efficacy could not be due to impaired absorption and inadequate plasma/tissue levels.

Another explanation for the lack of clinical efficacy of the compounds is that once the neurokinin-1 receptor binds substance P, it is internalized through a process of endocytosis, making the receptor unavailable to an extracellular antagonist [21]. This led to the hypothesis that these compounds may be more appropriate for the preventive treatment of migraine, prior to activation of the trigeminal nociceptive fibers. Unfortunately, a placebo-controlled trial with LY-303870 failed to demonstrate efficacy of the compound when used for migraine prevention [22].

Recent clinical studies with neurokinin-1-receptor antagonists have demonstrated their efficacy in treating depression and chemotherapy-induced nausea [23]. Whether they may ultimately be useful as anti-emetic agents during migraine headache is unclear. However, preliminary studies with LY-308870 and RPR-100893 in the abortive treatment of migraine failed to show efficacy for migraine-associated nausea, when compared to placebo.

The results of the above studies have clarified the lack of importance of substance P, acting peripherally or centrally *via* neurokinin-1 receptors, in the pathogenesis of the migraine headache, and call into question the overall relevance of plasma-protein extravasation.

Calcitonin gene-related peptide

Although stimulation of the trigeminal ganglion in animals and man results in the release of substance P and calcitonin gene-related peptide, only the latter is released after stimulation of pain-sensitive cranial structures, such as the superior sagittal sinus [24]. This finding raises the possibility that neurogenic vasodilation, mediated by calcitonin gene-related peptide, may be more relevant in the pathogenesis of the migraine headache than plasma-protein extravasation. A number of other observations appear to support this concept. During migraine headache, calcitonin gene-related peptide, and not substance P, is elevated in blood drawn from the external jugular vein [25]. Also, after administration of sumatriptan, the level of calcitonin gene-related peptide decreases with subsidence of the migraine headache [26]. Avitriptan, a potent serotonin-1B/D-receptor agonist with demonstrated efficacy in migraine, blocks the release of calcitonin gene-related peptide in animals but has much less effect on plasma-protein extravasation than sumatriptan [27]. Furthermore, CP-122288, a highly potent analogue of sumatriptan with no clinical antimigraine efficacy, powerfully inhibits neurogenic plasma-protein extravasation but is ineffective at blocking the release of calcitonin gene-related peptide [28, 29].

Additional evidence that calcitonin gene-related peptide plays a contributory role in the pathogenesis of migraine and other primary headaches comes from the observation that during spontaneous attacks of cluster headache and paroxysmal hemicrania, calcitonin gene-related peptide but not substance P is elevated in blood drawn from the external jugular vein [30, 31]. Furthermore, calcitonin gene-related peptide is increased in the peripheral circulation of migraineurs between headaches, which may reflect long-lasting (or permanent) impairment of neurogenic vascular control in these patients [32].

The above clinical observations support the relevance of recent anatomical findings. First, the number of cell bodies in the trigeminal ganglion of man that contain calcitonin gene-related peptide exceeds the number of those that contain substance P [33]. Electron microscopy has demonstrated the existence of vesicles immunoreactive to calcitonin gene-related peptide in perivascular nerve fibers of human superficial temporal artery [34]. The cerebral blood vessels of several animals, such as cat, guinea pig and rat, are innervated by trigeminal nerve fibers containing calcitonin gene-related peptide, which is released after stimulation of the ganglion. In addition, calcitonin gene-related peptide has been shown to produce sustained dilation of dural blood vessels in animals and, *in vitro*, of human middle meningeal artery [35].

The vasodilator effect of calcitonin gene-related peptide on isolated human cerebral artery is reversed by a specific receptor antagonist [36].

A recent double blind, crossover study evaluated the effect of human alpha-calcitonin gene-related peptide or placebo in 12 patients suffering from migraine without aura, infused intravenously with a dose of 2 μg/min for 20 min [37]. Within the subsequent 12 h, all patients experienced headache after infusion of alpha-calcitonin gene-related peptide, compared to only one patient after placebo infusion. In three patients, the headache fulfilled the criteria as proposed by the International Headache Society for migraine without aura. The mechanism of the migraine headache induced by alpha-calcitonin gene-related peptide is not known but may involve sensitization of trigeminal sensory nerve fibers or the release of nitric oxide [38]. It does not appear to be mediated through activation of potassium channels sensitive to adenosine triphosphate or calcium [39]. Whether inhibition of the release of calcitonin gene-related peptide or antagonism of its vasodilator effect will translate into acute headache relief in migraine patients is the focus of ongoing investigations.

Endothelins and other inhibitors of neurogenic inflammation

The endothelins represent potent endothelium-generated contractile peptides that are important mediators of endothelium-induced vasoconstriction. So far, three endothelin peptides have been identified (endothelin-1, -2, and -3) and two receptor subtypes have been cloned (endothelin-A, endothelin-1 selective; endothelin-B, endothelin non-selective). These receptors are present on human cerebral, meningeal and temporal arteries. Activation of endothelin-A and -B receptors leads to constriction and relaxation of cerebral blood vessels, respectively [40]. Endothelin-1, like calcitonin gene-related peptide, is increased during migraine headache [41], implying its involvement in the pathogenesis of the headache. However, bosentan, a non-specific endothelin-(A/B)-receptor antagonist without vasoconstrictor activity, was shown to be ineffective in aborting migraine headache, despite its ability to inhibit neurogenic inflammation [42].

The conclusion from this and other studies is that the ability of a compound to inhibit neurogenic inflammation in animals does not predict its therapeutic efficacy in migraine. The failure of potent inhibitors of neurogenic inflammation, such as neurokinin-1-receptor antagonists, 4991W93 (a selective inhibitor of neurogenic inflammation), bosetan and CP-122288 to abort migraine headaches, supports this conclusion [43, 44]. Indeed, the clinical failure of these potent inhibitors of neurogenic inflammation *and* the efficacy of compounds that are weak inhibitors of plasma-protein extravasation, such as alniditan, suggest that plasma-protein extravasation is unlikely to be a fundamental mechanism in the pathogenesis of the migraine headache. Furthermore, the results indicate that inhibition of neurogenic inflammation in animals is neither necessary nor sufficient for clinical efficacy. This assertion is underscored by the failure of magnetic resonance imaging to demonstrate the phenomenon of plasma-protein extravasation in man [45], as well as the failure to demonstrate this phenomenon in the retina of patients during migraine or cluster headache [46]. Magnetic resonance imaging is, however, not the most sensitive technique to demonstrate the occurrence of neurogenic inflammation. A study using single photon-emission computed tomography of the brain with radioactive labeled albumin provided preliminary evidence for the occurrence of plasma-protein extravasation extraparenchymally, during and ipsilateral to migraine headache [47].

Conclusion

Activation of trigeminal nociceptive nerve fibers in the rat produces a perivascular response, plasma-protein extravasation, and vasodilation. Plasma-protein extravasation and vasodilation are mediated by substance P and calcitonin gene-related peptide, respectively. Both substance P and calcitonin gene-related peptide are present in the trigeminal nociceptive nerve fibers in man and are released after electrical stimulation of the trigeminal ganglion. However, only calcitonin gene-related peptide is released after stimulation of pain-sensitive cranial structures in animals and during migraine and related headaches. In keeping with these observations, the phenomenon of plasma-protein extravasation does not appear to be relevant in the pathogenesis of the migraine headache. Thus far, despite their potent inhibition of neurogenic inflammation, endothelin-1- and neurokinin-1-receptor antagonists, as well as CP-122288 and 4991W93, have been disappointing therapeutic failures in treating migraine headache. Moreover, avitriptan and BMS-181885, which are clinically effective antimigraine medications, are 400- and 900-fold less potent, respectively, than sumatriptan in reducing plasma-protein extravasation after stimulation of the trigeminal ganglion.

The role of neurogenic vasodilation, which is mediated by the release of calcitonin gene-related peptide from the nociceptive nerve terminals, is unclear at this time. However, inhibition of the release of calcitonin gene-related peptide is a promising therapeutic target, given the significant animal and human data that support a role for this neuropeptide in the pathogenesis of the migraine headache.

References

1 Lipton RB, Stewart WF (1993) Migraine in the United States: epidemiology and health-care utilization. *Neurology* 43: 6–10

2 Goadsby PJ, Zagami AS, Lambert GA (1991) Neural processing of craniovascular pain: a synthesis of the central structures involved in migraine. *Headache* 31: 365–371

3 Wolff HG (1963) Headache and other head pain. Oxford University Press, New York
4 Mayberg MR, Zervas NT, Moskowitz MA (1984) Trigeminal projections to supratentorial pial and dural blood vessels in cats demonstrated by horseradish peroxidase histochemistry. *J Comparative Neur* 223: 46–56
5 Mayberg MR, Langer RS, Zervas NT, Moskowitz MA (1981) Perivascular meningeal projections from cat trigeminal ganglia: possible pathway for vascular headaches in man. *Science* 213: 228–230
6 Dimitriadou V, Buzzi MG, Theoharides TC, Moskowitz MA (1992) Ultrastructural evidence for neurogenically mediated changes in blood vessels of the rat dura mater and tongue following antidromic trigeminal stimulation. *Neuroscience* 48: 187–203
7 Markowitz S, Saito K, Moskowitz MA (1987) Neurogenically mediated leakage of plasma proteins occurs from blood vessels in the dura mater but not brain. *J Neurosci* 7: 4129–4136
8 Dimitriadou V, Buzzi MG, Moskowitz MA (1991) Trigeminal sensory fiber stimulation induces morphological changes in rat dura mater mast cells. *Neuroscience* 44: 97–112
9 Edvinsson L, Hara H, Uddman R (1989) Retrograde tracing of nerve fibers to the rat middle cerebral artery with true blue: co-localization with different peptides. *J Cerebral Blood Flow Metabol* 9: 212–218
10 Saria AL, Lundberg J, Skofitsch G et al (1983) Vascular leakage in various tissues induced by substance P, capsaicin, bradykinin, serotonin, histamine and antigen challenge. *Naunyn Schmiedebergs Arch Parmacol* 324: 212–218
11 Cutrer FM, Limroth V, Waeber C et al (1997) New targets for antimigraine drug development. In: PJ Goadsby, SD Silberstein (eds): *Headache*. Butterworth-Heinemann, Boston, 59–72
12 Goadsby PJ, Edvinsson L, Ekman R (1988) Release of vasoactive peptides in the extracerebral circulation of humans and the cat during activation of the trigeminovascular system. *Ann Neurol* 23: 193–196
13 Goadsby PJ, Edvinsson L (1993) The trigeminovascular system and migraine: studies characterising cerebrovascular and neuropeptide changes seen in man and cat. *Ann Neurol* 33: 48–56
14 Edvinnson L, Goadsby PJ (1998) Neuropeptides in headache. *Eur J Neurol* 5: 329–341
15 Cutrer FM, Moussaoui S, Garret C, Moskowitz MA (1995) The non-peptide neurokinin-1 antagonist, RPR 100893, decreases c-fos expression in trigeminal nucleus caudalis following noxious chemical meningeal stimulation. *Neuroscience* 64: 741–750
16 Shepheard SL, Williamson DJ, Williams J et al (1995) Comparison of the effects of sumatriptan and the NK-1 antagonist CP-99,994 on plasma extravasation in dura mater and c-fos mRNA expression in trigeminal nucleus caudalis of rats. *Neuropharmacol* 34: 255–261
17 Shepheard SL, Williamson DJ, Hill RG, Hargreaves RJ (1993) The non-peptide NK-1 receptor antagonist RP 67580 blocks neurogenic plasma extravasation in the dura mater of rats. *Br J Pharmacol* 108: 11–12
18 Gitter BD, Bruns RF, Howbert JJ et al (1995) Pharmacological characterization of

LY303870: a novel, potent, and selective nonpeptide substance P (neurokinin 1) receptor antagonist. *J Pharmacol Exp Ther* 275: 737–744

19 Goldstein DJ, Wang O, Saper JR (1997) Ineffectiveness of NK-1 antagonist in acute migraine: a crossover study. *Cephalalgia* 17: 785–790

20 Connor HE, Bertin L, Gillies S et al (1998) Clinical evaluation of a novel, potent, CNS penetrating NK1 receptor antagonist in the acute treatment of migraine (abstract). *Cephalalgia* 18: 293

21 Mantyh PW, DeMaster E, Malhotra A et al (1995) Receptor endocytosis and dendrite reshaping in spinal neurons after somatosensory stimulation. *Science* 268: 1629–1632

22 Goldstein DJ, Offen WW, Klein EG, Phebus LA (1999) Lanepitant, an NK-1 antagonist, in migraine prophylaxis (abstract). *Cephalalgia* 19: 377

23 Saria A (1999) The tachykinin NK_1 receptor in the brain: pharmacology and putative functions. *Eur J Pharmacol* 375: 51–60

24 Zagami AS, Goadsby PJ, Edvinsson L (1990) Stimulation of the superior sagittal sinus in the cat causes release of vasoactive peptides. *Neuropeptides* 16: 69–75

25 Goadsby PJ, Edvinsson L, Ekman R (1990) Vasoactive peptide release in the extracerebral circulation of humans during migraine headache. *Ann Neurol* 28: 183–187

26 Goadsby PJ, Edvinsson L (1993) The trigeminovascular system and migraine: studies characterizing cerebrovascular and neuropeptide changes seen in man and cat. *Ann Neurol* 33: 48–56

27 Couch JR, Saper J, Meloche JP (1996) Treatment of migraine with BMS 180048: response at 2 hours. *Headache* 36: 523–530

28 Knight YE, Edvinsson L, Goadsby PJ (1997). Blockade of release of CGRP after superior sagittal sinus stimulation in cat: a comparison of avitriptan and CP122,288 (abstract). *Cephalalgia* 17: 248

29 Roon K, Diener HC, Ellis P et al (1997) CP-122,288 blocks neurogenic inflammation, but is not effective in aborting migraine attacks: results of two controlled clinical trials (abstract). *Cephalalgia* 17: 245

30 Fanciullaci M, Alessandri M, Fegini M et al (1995) Increases in plasma calcitonin gene-related peptide from extracerebral circulation during nitroglycerin-induced cluster headache attack. *Pain* 60: 119–123

31 Goadsby PJ, Edvinsson L (1994) Human *in vivo* evidence for trigeminovascular activation in cluster headache. *Brain* 117: 427–434

32 Ashina M, Bendtsen L, Jensen R et al (1999) Increased plasma levels of calcitonin gene-related peptide in migraine (abstract). *Cephalalgia* 19: 346

33 Tajti J, Uddman R, Møller F et al (1999) Neuronal messenger molecules and peptide receptors distribution in the human trigeminal ganglion (abstract). *Cephalalgia* 19: 392

34 Van Rossum D, Hanisch UK, Quirion R (1997) Neuroanatomical localization, pharmacologic characterization and functions of CGRP, related peptides and their receptors. *Neurosci Biobehav Rev* 21: 649–678

35 Jansen-Olesen I, Mortensen A, Edvinsson L (1996) Calcitonin gene-related peptide is

released from capsaicin-sensitive nerve fibers and induces vasodilation of human cerebral arteries concomitant with activation of adenylyl cyclase. *Cephalalgia* 16: 310–316
36 Moreno MJ, Abounader R, Lantin S et al (2000) CGRP-mediated dilations in human cerebral arteries are potently antagonized by BIBN4096BS, a selective CGRP receptor antagonist (abstract). *Cephalalgia* 20: 272–273
37 Lassen LH, Jacobson VB, Pedersen PA et al (1999) hCGRP-induced headache in migraineurs (abstract). *Cephalalgia* 19: 316
38 De Hoon JNJM, Pikkers P, Smits P et al (1999) Calcitonin gene-related peptide induces vasodilation by nitric oxide (abstract). *Cephalalgia* 19: 315
39 De Hoon JNJM, Pikkers P, Smits P et al (1999) Calcitonin gene-related peptide-induced vasodilation does not involve the activation of potassium channels in man (abstract). *Cephalalgia* 19: 397
40 Jansen-Olesen I, Edvinnson L (2000) Cephalic neurovascular transmitters and receptors. In: J Olesen, P Tfelt-Hansen, KMA Welch (eds): *The headaches.* Raven Press, New York, 87–98
41 Hasselblatt M, Köhler J, Volles E, Ehrenreich H (1999) Simultaneous monitoring of endothelin-1 and vasopressin plasma levels in migraine. *Neuroreport* 10: 423–425
42 May A, Gijsman HJ, Wallnofer A et al (1996) Endothelin antagonist bosentan blocks neurogenic inflammation, but is not effective in aborting migraine attacks. *Pain* 67: 375–378
43 Earl NL, McDonald SA, Lowy MT (1999) Efficacy and tolerability of the neurogenic inflammation inhibitor, 4991W93, in the acute treatment of migraine (abstract). *Cephalalgia* 19: 357
44 Roon KI, Olesen J, Diener HC et al (2000) No acute antimigraine efficacy of CPP-122,288, a highly potent inhibitor of neurogenic inflammation: results of two randomized, double blind, placebo-controlled clinical trials. *Ann Neurol* 47: 238–241
45 Nissila M, Parkkola R, Sonninen P, Salonen R (1996) Intracerebral arteries and gadolinium enhancement in migraine without aura (abstract). *Cephalalgia* 16: 363
46 May A, Shepheard S, Wessing A et al (1998) Retinal plasma extravasation can be evoked by trigeminal stimulation in rat but does not occur during migraine attacks. *Brain* 121: 1231–1237
47 Pappagalo M, Szabo Z, Esposito G et al (1999) Imaging neurogenic inflammation in patients with migraine headaches (abstract). *Neurology* 52: A274–A275

Models of neurogenic inflammation as it relates to migraine

Uwe Reuter and Guy Arnold

Humboldt Universität zu Berlin, Charité, Neurologische Klinik und Poliklinik, Schumannstrasse 20–21, 10098 Berlin, Germany

Introduction

For more than half a century, migraine was believed to be a vascular disorder with the aura attributed to cerebral vasoconstriction and ischemia and the headache to dilation and inflammation of extracranial arteries. However, there is mounting evidence that cortical spreading depression is the mechanism underlying the migraine aura and that (neurogenic) inflammation is more important than vasodilation in the migraine headache. This chapter focuses on the experimental models related to the putative migraine mechanisms of cortical spreading depression and neurogenic inflammation (Tab. 1).

Neurogenic inflammation in the dura mater

The pain of the migraine headache is not only transmitted predominantly by trigeminal nerve fibers, which originate from cell bodies in the trigeminal ganglion, but also involves their active participation [1]. In particular the nerve fibers in the first branch of the trigeminal nerve, the ophthalmic nerve, are involved, which innervate the forehead and anterior vertex. These nerve fibers also innervate a large part of the dura mater, the connective tissue membrane lying against the inside of the skull, which has served as a useful location to study the effects of trigeminal activation [2, 3]. It is in relation to the dura mater that, in actual fact, most of the research related to neurogenic inflammation has been conducted. In addition, the specific abortive antimigraine medications, the ergots and triptans, have been found to inhibit neurogenic inflammation in the dura mater but not in the extracranial tissues.

Neurogenic inflammation is mediated by the release of peptides from the sensory nociceptive nerve fibers. These peptides, so-called neuropeptides, are calcitonin gene-related peptide, substance P and neurokinin A [4–6]. The inflammation occurs

Migraine: A Neuroinflammatory Disease?, edited by Egilius L.H. Spierings and Margarita Sánchez del Río

Table 1 - Experimental models used in relation to migraine

Model	Refs.	Stimulus	Objective	Species
Trigeminal ganglion stimulation	[9]	electrical	protein extravasation	guinea pig
	[120]		neuropeptide release	rat
	[45]		regional cerebral blood flow	mouse
	[121]		c-fos expression	guinea pig
	[10]			
	[9]	chemical	protein extravasation	guinea pig
	[122]		c-fos expression	rat
	[9]	immunological	protein extravasation	guinea pig
Sphenopalatine ganglion stimulation	[70]	electrical	protein extravasation	rat
Superior sagittal sinus stimulation	[82]	electrical	neuropeptide release	cat
	[46]	mechanical	neuronal activity	
	[83]		regional cerebral blood flow	
	[84]		c-fos expression	
	[15]	chemical	neuropeptide release	guinea pig
Dura stimulation	[80]	electrical	vessel diameter	cat
	[79]		neuronal activity	rat
	[40]			
	[79]	mechanical	neuronal activation and	rat
	[76]		sensitization	
			c-fos expression	
	[15]	chemical	neuropeptide release	rat
	[23]		c-fos expression	guinea pig
	[24, 25]			
Cortical spreading depression	[105]	electrical	electrophysiology	rabbit
	[107]	mechanical	cerebral blood flow	rat
	[113]	chemical	c-fos expression	
Genetic mutants	[123]	chemical	electrophysiology	mouse
Isolated blood vessels	[124]	chemical	blood vessel diameter	man
	[96]			

in the peripheral tissues and not in the central nervous system, although the same peptides are released there as neurotransmitters. The peptides have their effects on vascular smooth muscle and endothelial cells as well as on mast cells and blood platelets, as documented by histological and electron microscopic studies [7, 8]. It is the activation of the endothelial cells that, through increased transendothelial transport, results in the extravasation of plasma proteins, the focus of most studies.

Stimulation of the trigeminal ganglion

In 1987, Markowitz et al. observed that (retrograde) electrical stimulation of the trigeminal ganglion resulted in a marked increase in plasma-protein extravasation in the dura mater, ipsilateral to the side of stimulation [9]. This was also observed in the ipsilateral extracranial tissues but not in the brain, possibly due to a different structure of the cerebral blood vessels. These blood vessels, in contrast to those elsewhere in the body, have tight endothelial junctions, which form the blood-brain barrier and create a physical barrier, possibly also against plasma-protein extravasation. The initial experiments were performed on anaesthetized rats but studies have since also been conducted on guinea pigs and mice with similar results [2, 10].

In the experiments, bovine serum albumin labeled with 125-iodine or Evans blue, labeling the plasma proteins present in the body, is used to quantify the plasma-protein extravasation. After injection of the iodine-labeled albumin or Evans blue into the femoral vein, the anesthetized animal is fixed in a stereotaxic frame. Its skull is exposed and burr holes are drilled symmetrically in the skull above the trigeminal ganglia, using stereotactic coordinates. Paired non-concentric bipolar electrodes are placed through the burr holes in the trigeminal ganglia to be stimulated for several minutes, starting 5 min after the injection of the iodine-labeled albumen or Evans blue. In the initial experiments, the stimulation occurred for 5 min with paired rectangular pulses of opposite polarity, at a frequency of 5 Hz, duration of 5 msec, and intensity of 3 mA.

In subsequent experiments, the time of stimulation, as well as the frequency and duration of the stimuli, was kept constant while the intensity was changed. However, no formal dose-response studies have been done to determine the effects of stimulation with different intensities on plasma-protein extravasation. After termination of the stimulation, the animal is perfused transcardially with 0.9% sodium chloride. The dura mater is then carefully removed on both sides of the skull and the amount of extravasated plasma proteins, labeled with 125-iodine or Evans blue, determined. The results are expressed as a ratio of radioactive counts or light absorption, respectively, per microgram of dural tissue on the stimulated *versus* non-stimulated side. Spectrophotofluorometry or colorimetry at the absorption maximum of 620 nm is used to measure Evans blue, which is a less sensitive technique, however, than the use of a radioactive tracer [9, 11].

With the above method, the plasma-protein extravasation occurs within minutes after stimulation and continues for hours thereafter, even following short stimulation. Using a typical stimulation pattern in the rat (5 min; 5 Hz; 5 ms; 0.6 mA), Huang et al. showed that plasma-protein extravasation can be detected for up to 2 h after stimulation [12]. In the study, the tracer was administered 5 min before the stimulation and the animals were sacrificed immediately, 30, 60, 90, 120, or 180 min later. The ratio of stimulated *versus* non-stimulated side peaked 5 min after the stimulation at approximately 1.7 and, beginning 60 min later, slowly decreased to 1.25 at 2 h and 1 at 3 h. The study also showed that plasma proteins accumulating in the dura mater during and after trigeminal stimulation remain present for at least 2 h.

In order to determine whether plasma-protein extravasation occurs during stimulation only, iodine-labeled albumin was administered immediately and at time points 0 to 120 min after stimulation. The plasma-protein extravasation was found to be significantly increased on the stimulated *versus* non-stimulated side with the iodine-labeled albumin administered up to 90 min after stimulation. However, if the injection occurred 2 h after stimulation, it was not associated with a difference in plasma-protein extravasation in the dura mater between the two sides. It can, therefore, be concluded that plasma-protein extravasation due to stimulation of the trigeminal ganglion occurs for up to 90 min after stimulation.

Other methods of inducing neurogenic inflammation

Williamson et al. applied the electrical stimulation not to the trigeminal ganglion but directly to the exposed dura mater and measured blood-vessel diameter with the use of intravital microscopy to study neurogenic vasodilation [13]. Other than with electrical stimulation, plasma-protein extravasation can be induced with chemicals administered intravenously or intrathecally, such as capsaicin or substance P, or with immunological stimuli. A disadvantage of these particular methods vis-à-vis stimulation of the trigeminal ganglion is the inability to create side differences, as possibly occur during unilateral migraine headaches. The ability to compare physiology and pathology within the same animal, as is possible with the stimulation of the trigeminal ganglion, might more closely mimic the situation as it occurs during migraine headache.

Initially described by Gepetti et al., capsaicin depolarizes perivascular nerve fibers, releasing neuropeptides from them, which, in turn, cause plasma-protein extravasation mostly at the level of the post-capillary venules [14, 15]. Plasma-protein extravasation after electrical stimulation of the trigeminal ganglion is markedly reduced in animals treated as neonates with capsaicin. Capsaicin, the extract of red pepper, destroys the C-fibers when administered neonatally [16]. In such animals pretreated with capsaicin, the effect of substance P in causing plasma-protein extravasation is unchanged compared to controls [17]. This is because substance P acts directly on the post-capillary venules, causing plasma-protein extravasation through a non-neurogenic mechanism. It can, therefore, be concluded that neurogenic plasma-protein extravasation is mediated, at least in part, by activation of C-fibers [18].

Non-neurogenic plasma-protein extravasation, not dependent on the release of neuropeptides, can also be induced by intravenous administration of serotonin, bradykinin, or histamine [9, 14]. Non-neurogenic plasma-protein extravasation is dose-dependent, in contrast to neurogenic plasma-protein extravasation, for example, induced by capsaicin, which is not. A threshold dose of capsaicin of 1 µmol/kg causes a 25% increase in plasma-protein extravasation, while a dose of 3 µmol/kg is usually lethal. This narrow dose range may not allow the establishment of a dose-response relationship. On the other hand, it is also possible that C-fibers have an all-or-nothing response to capsaicin, thus causing plasma-protein extravasation after capsaicin administration, independent of the administered dose [9, 19, 20].

Immunological activation of the trigeminal system is rarely used to study neurogenic inflammation. It involves sensitization of the animal with an intra-abdominal injection of ovalbumin. A second dose of ovalbumin, administered intravenously 4–6 weeks later, induces plasma-protein extravasation that can be inhibited by pretreatment with a combination of histamine-1- and -2-receptor antagonists. This suggests an important role of histamine and, therefore, of mast cells, the predominant source of histamine in the body, in immunologically induced plasma-protein extravasation [9]. The role of mast cells in plasma-protein extravasation induced otherwise has not been clearly determined yet, although [14] several studies have shown activation of mast cells to occur in neurogenic inflammation [7, 8, 14, 21]. Determination of the precise role that mast cells play in neurogenic inflammation has been hampered by the lack of "complete" mast cell inhibitors, the great variety of granules they contain, and their complex interaction with other cells. A possible approach to further study this issue is by means of genetically manipulated mice or rats that are deficient in mast cells.

Electrical stimulation of the dura mater or trigeminal ganglion activates second-order neurons in the trigeminal nucleus caudalis, in the same way as chemical stimulation of the dura mater does. Intravenous administration of capsaicin causes plasma-protein extravasation in the dura mater [9, 13] and so does injection of microquantities of the compound into the trigeminal ganglion (MA Moskowitz; personal communication). Capsaicin also causes expression of c-fos in the dorsal horn of the spinal cord of neonatal rat [22], c-fos being a gene expressed immediately upon stimulation of a pain-sensitive structure. When administered intracisternally, capsaicin, as well as blood, induces the expression of c-fos in the trigeminal nucleus caudalis of the rat and guinea pig [23–25].

C-fos is also expressed in the trigeminal nucleus caudalis after systemic administration of nitroglycerin [26]. The neurons positive for c-fos have been found to be in close proximity to nerve fibers showing increased activity of nitric oxide synthase [27]. Thus, increased production of nitric oxide occurs after administration of nitroglycerin in close proximity to neurons positive for c-fos in the trigeminal nucleus caudalis. This suggests a possible involvement of nitric oxide in the activation of nociceptive neurons in the trigeminal nucleus caudalis after administration of nitroglycerin. Administration of nitroglycerin in man causes immediate headache and, only in migraineurs, delayed headache resembling migraine headache [28]. In addition, administration of the inhibitor of nitric oxide synthase, l-n^{G}-methylarginin, improves or abolishes migraine headache [29]. Subcutaneous administration of nitroglycerin to rats also caused nitric oxide synthase to appear in an increased number of nerve fibers in the dura mater. It also caused structural changes of these fibers to occur in terms of beading, further reflecting activation of the nerve fibers upon administration of nitroglycerin [30].

Inhibition of neurogenic inflammation

Neurogenic plasma-protein extravasation in the dura mater, but not in the extracranial tissues, can be blocked by pre-treatment with ergots in clinically relevant doses [31]. Non-neurogenic plasma-protein extravasation caused by the administration of substance P or neurokinin A is not affected, which implies a presynaptic rather than vascular mechanism. In 1991, Buzzi and Moskowitz discovered that the first triptan, sumatriptan, a selective serotonin-1B/D-receptor agonist and effective abortive antimigraine medication, inhibits neurogenic inflammation [18]. This led to the discovery of serotonin receptors on the trigeminal nerve fibers in the dura mater that are involved in neurogenic inflammation [32, 33]. The serotonin receptors have a function in modulating neurogenic inflammation, providing a model for the study of the effects of antimigraine medications. Serotonin receptors, mainly of the 1B type, are also present on cerebral blood vessels [34]. Sumatriptan, as well as the ergots, constricts these blood vessels *in vitro* and they do the same with the isolated human middle meningeal and temporal arteries. Sumatriptan causes vasoconstriction by stimulating vascular serotonin-1B receptors with little, if any, contribution of the serotonin-1F receptor [35]. Sumatriptan, as well as the other triptans, reduces c-fos expression in the trigeminal nucleus caudalis of the rat induced by noxious stimulation [23]. Therefore, a purely vascular mode of action is not likely to fully explain the beneficial effect of the medication in the abortive treatment of migraine.

The serotonin-1F-receptor agonist, LY-334370, which lacks vasoconstrictor activity [36], is effective in the abortive treatment of migraine but in high doses only. It has no effect on neurogenic *vasodilation* in the dura mater of the rat, caused by electrical stimulation of the dura, but successfully inhibits neurogenic inflammation [37]. Clinical studies with antagonists of the neurokinin-1-receptor have failed to show benefit in migraine abortion, while they effectively block plasma-protein extravasation [38].

Specific antimigraine medications, such as the ergots and triptans, as well as the non-steroidal anti-inflammatory analgesics, significantly reduce neurogenic inflammation in animals [18, 19, 39–41]. Aspirin and indomethacin are though to act pre- as well as postjunctionally to block plasma-protein extravasation. Indomethacin blocks plasma-protein extravasation in the dura mater of the rat induced by electrical stimulation of the trigeminal ganglion. It does so at the low dose of 1 mg/kg, which does not affect the plasma-protein extravasation induced by substance P. In contrast, the latter can be inhibited only by a 10-fold higher dose of the medication (10 mg/kg). This suggests that non-neurogenic plasma-protein extravasation is blocked by an action on the vascular changes induced by the neuropeptides. Since neurogenic inflammation is inhibited by a much lower dose of indomethacin, it is likely that the non-steroidal anti-inflammatory analgesics also act prejunctionally by inhibiting the release of neuropeptides. Dexamethasone, on the other hand, used to abort migraine status [42, 43], does not block neurogenic inflammation in animals, suggesting a different mode of action [41].

Neuropeptides

In 1984, it was found that cultured neurons of the trigeminal ganglion, upon stimulation, release calcitonin gene-related peptide [44]. Several studies have since shown neuropeptide release to occur in animals upon stimulation of the trigeminal ganglion. In the rat and cat, stimulation of the trigeminal ganglion causes release of neuropeptides from C-fibers in the dura mater, as observed in blood drawn from the external jugular vein or superior sagittal sinus [31, 45]. Electrical stimulation of the trigeminal ganglion in the rat increases the level of calcitonin gene-related peptide in blood drawn from the superior sagittal sinus, dependent on the intensity of stimulation (0.1–1.0 mA). The level of calcitonin gene-related peptide increases 7–8-fold, starting at a baseline level of 2 pmol/l, ultimately reaching 17 pmol/l following trigeminal stimulation with an electric current of 1 mA. The level of calcitonin gene-related peptide in the superior sagittal sinus is maximal after stimulation for 1 min and does not increase further, despite ongoing stimulation for up to 3 min. The level of the peptide starts to decline slowly after that but is still slightly elevated 10 min later. The level of calcitonin gene-related peptide in the systemic circulation, determined in blood drawn from a femoral artery, does not change after trigeminal stimulation, which may be due to dilution or clearance of the peptide. Pre-treatment with sumatriptan or an ergot in doses that inhibit neurogenic plasma-protein extravasation blocks the increase in calcitonin gene-related peptide in blood drawn from the superior sagittal sinus after trigeminal stimulation, thus indicating a prejunctional mode of action.

Direct stimulation of the superior sagittal sinus, a structure densely innervated by sensory nerve fibers, also results in elevated levels of calcitonin gene-related peptide in blood drawn from an extracerebral vein [46]. Electrical stimulation of the sagittal sinus in the cat leads to an increase of 85% in the level of calcitonin gene-related peptide in jugular vein blood. It also leads to an increase of 300% in the level of vasoactive intestinal polypeptide, a major neurotransmitter of the parasympathetic nervous system. However, the level of substance P does not change after electrical stimulation of the sagittal sinus, possibly due to rapid degradation of this peptide in the plasma.

The release of neuropeptides has also been studied in an *in vitro* preparation of the fluid filled skull cavity of the rat. Electrical stimulation of the trigeminal ganglion or chemical stimulation of the dural nerve fibers with a combination of inflammatory mediators (serotonin, histamine and bradykinin) cause neuropeptide release. Electrical stimulation of the trigeminal ganglion (5 min) increased the level of calcitonin gene-related peptide in the fluid by 4–5-fold, returning to baseline as soon as the stimulation was terminated. The level of substance P in the skull-cavity fluid was not affected by the electrical stimulation [47]. It is interesting that the level of prostaglandin E2 also significantly increased in the fluid after electrical stimulation of the trigeminal ganglion or chemical stimulation of the dural nerve fibers. Prostaglandins have been proposed to play a role in neurogenic inflammation as well as in the pathogenesis of the migraine headache.

Based on experiments in neuronal cultures of the trigeminal ganglion, we have advanced our understanding of the mechanisms by which stimulated release of neuropeptides can be inhibited. Serotonin agonists, such as sumatriptan, bind to serotonin-1B and -D receptors on trigeminal nerve fibers but – in contrast to the classical effect ascribed to stimulation of serotonin receptors – increase intracellular calcium instead of reducing intracellular cyclic AMP. Furthermore, a phosphatase inhibitor effectively blocked the inhibitory effect of sumatriptan in cell cultures on the stimulated release of calcitonin gene-related peptide [48]. This suggests that sumatriptan causes an increase in phosphatase activity *via* a calcium-dependent mechanism.

In man, electrocoagulation of the trigeminal ganglion results in increased levels of calcitonin gene-related peptide and substance P in blood drawn from the external jugular but not the antecubital vein [45]. This suggests that stimulation of the trigeminal ganglion in man also causes the peripheral release of neuropeptides. It is interesting that elevated levels of calcitonin gene-related peptide were observed during migraine headache in blood drawn from the external jugular vein, suggesting that trigeminal sensory fibers are, indeed, activated during migraine [49]. Treatment of the headache with sumatriptan not only aborted the pain but also reduced the elevated levels of this peptide [50].

In animals, it has been observed that the number of neurons containing calcitonin gene-related peptide, substance P and neurokinin A in the trigeminal nucleus caudalis markedly decreases after trigeminal stimulation [51, 52]. This finding suggests a central release of neuropeptides as well, acting as co-transmitters or co-modulators in the trigeminal sensory system. It is surprising that substance P, the neuropeptide thought to be primarily involved in plasma-protein extravasation, was not found to be elevated in blood drawn from the external jugular vein during migraine headache. It is not clear why this is so, but it could be due to a lesser abundance in trigeminal nerve fibers of substance P in comparison to calcitonin gene-related peptide, a lower sensitivity of the assay, dilution of the peptide, or rapid metabolism. However, it may also suggest that substance P – in contrast to calcitonin gene-related peptide – is not involved in the pathogenesis of the migraine headache.

Although the neuropeptides are important in neurogenic inflammation, it is questionable whether calcitonin gene-related peptide and substance P are the *primary* chemical mediators in the development of the migraine headache. Injection of substance P or calcitonin gene-related peptide into the temporalis muscle did not cause pain [53] and intravenous administration of calcitonin gene-related peptide did not induce headache [54]. This may lead to the conclusion that other chemical mediators, such as nitric oxide, for example, are involved in the generation of the migraine headache.

Not much is known about the relationship between the neuropeptides, serotonin, and the production of nitric oxide. Serotonin, a key neurotransmitter involved in migraine, stimulates serotonin-1B and -2B receptors, causing increased synthesis of nitric oxide by endothelial cells [55, 56]. This increased synthesis of nitric oxide can be inhibited by methiothepin, a non-selective serotonin-1/2-receptor antagonist. Furthermore, *in vitro* studies have shown that serotonin-1B/D-receptor agonists inhibit the release of nitric oxide from brain slices [57]. Although there is evidence for synthesis of nitric oxide mediated by activation of serotonin receptors, the precise role of nitric oxide in neurogenic inflammation remains to be determined. Nitric oxide donors, such as nitroglycerin and sodium nitroprusside, cause pial vasodilation in the cat, which is almost completely abolished by trigeminal denervation or administration of the receptor antagonist for calcitonin gene-related peptide, calcitonin gene-related peptide 8-37 [58]. Thus, it appears that nitric oxide activates trigeminal nerve fibers to release calcitonin gene-related peptide, which, in turn, causes the dilation of the pial arteries.

Stimulation of the sphenopalatine ganglion

Several recent studies have linked the parasympathetic nervous system as well to neurogenic inflammation in the dura mater [59, 60]. Since γ-aminobutyric acid (GABA) is an inhibitory neurotransmitter and modulates the activity of both the trigeminal and parasympathetic systems, this section will also include a discussion of the effects of GABA-like compounds.

Glutamic acid decarboxylase, a GABA synthesizing enzyme, is found co-localized with calcitonin gene-related peptide in nerve fibers in the dura mater. However, it is not found co-localized with the parasympathetic neurotransmitters, acetylcholine, vasoactive intestinal polypeptide, and neuropeptide Y. Glutamic acid decarboxylase is also present in the trigeminal ganglion, some of it co-localized with calcitonin gene-related peptide, suggesting an interaction between this peptide and GABA [61].

Valproic acid, a preventive antimigraine medication, increases the GABA level by inhibiting the GABA metabolizing enzyme (GABA-transaminase) and stimulating its synthesizing enzyme, glutaminic acid decarboxylase. Intraperitoneal injection of valproic acid or the $GABA_A$-receptor agonist, muscimol, dose-dependently inhibited plasma-protein extravasation in the rat [62]. This occurred with plasma-protein extravasation induced by trigeminal stimulation (5 min; 5 Hz; 5 ms; 0.6 mA) or administration of substance P. The inhibitory effect was, in turn, completely reversed by pre-treatment with the $GABA_A$-receptor antagonist, bicuculline, whereas the B-receptor antagonist, baclofen, was not effective. Neonatal treatment with capsaicin did not affect the inhibitory effect of valproic acid on plasma-protein extravasation induced by substance P. In another series of experiments, the effect of the neurosteroid, progesterone, on plasma-protein extravasation was inhibited by the $GABA_A$-receptor agonists, bicuculline and bicuculline methiodide. Bicuculline methiodide, in contrast to bicuculline, does not cross the blood-brain barrier. Therefore, plasma-protein extravasation is mediated, at least in part, by $GABA_A$ receptors, located outside the blood-brain barrier. The $GABA_A$ receptor mediates c-fos expression in the trigeminal nucleus caudalis after intracisternal injection of capsaicin, which can be blocked by valproic acid. The $GABA_A$-receptor antagonists, bicuculline and allopregnanolone, can, in turn, antagonize the blockade caused by valproic acid.

Edema in peripheral tissues induced by calcitonin gene-related peptide can also be completely blocked by intrathecal injection of the $GABA_A$-receptor antagonist, bicuculline [63]. Furthermore, the potassium-evoked release of calcitonin gene-related peptide in dorsal root ganglia slices can be partially blocked by GABA and muscimol but not by baclofen. By blocking $GABA_A$-receptors, bicuculline methiodide completely prevents the blockade by GABA and muscimol. In contrast, neither GABA nor muscimol or baclofen alter the potassium-evoked release of substance P. These observations suggest that GABA exerts prejunctional inhibitory control on the primary sensory nerve fibers in the dorsal horn of the rat spinal cord that contain calcitonin gene-related peptide [64].

Our knowledge about parasympathetic cranial ganglia, the sphenopalatine ganglion, for example, and GABA is very limited. We know from experiments on pelvic and intestinal (parasympathetic) ganglia in the cat that they contain receptors for this particular inhibitory neurotransmitter. Activation of the $GABA_A$ receptor resulted in a biphasic depolarization of the ganglia [65] and caused, at least in intestinal ganglia, the release of acetylcholine [66]. Moreover, it has been demonstrated that substance P depolarizes parasympathetic ganglia [67], which can be blocked by GABA, indicating a possible connection between neuropeptides, GABA and the parasympathetic nervous system.

The human otic and sphenopalatine ganglia – the origin of the parasympathetic nerve fibers that innervate the dural blood vessels – contain cell bodies positive for nitric-oxide synthase and vasoactive intestinal polypeptide, as well as for messenger RNA encoding for the receptors for vasoactive intestinal polypeptide and neuropeptide Y-1. Receptors for calcitonin gene-related peptide are not present in these parasympathetic ganglia [68]. In contrast, human trigeminal ganglia contain messenger RNA encoding for the receptors for calcitonin gene-related peptide 1, neuropeptide Y-1, neuropeptide Y-2 and vasoactive intestinal polypeptide 1, indicating that the activity of the trigeminal system is, in part, under parasympathetic control [69].

Electrical stimulation of the sphenopalatine ganglion also causes plasma-protein extravasation in the ipsilateral dura mater, with stimulation parameters similar to those for stimulation of the trigeminal ganglion. Plasma-protein extravasation after sphenopalatine stimulation can be completely blocked by the muscarinic-receptor antagonist, atropine [70], and is significantly reduced by pre-treatment with capsaicin. The latter indicates that stimulation of the sphenopalatine ganglion also leads to activation of sensory nerve fibers of the C-type, most likely belonging to the trigeminal system.

The role of the parasympathetic nervous system is also far from clear when it comes to the pathogenesis of migraine.

In chronic paroxysmal hemicrania, calcitonin gene-related peptide and vasoactive intestinal polypeptide, a neuropeptide widely distributed in parasympathetic nerve fibers, have been found to be elevated during the presence of headache in blood drawn from the external jugular vein [71]. This finding suggests a role of the parasympathetic system, in addition to the trigeminal system, in the so-called short-lasting headaches. In addition, it raises the question whether plasma-protein extravasation, mediated at least in part by these neurotransmitters, is also an important mechanism in headaches other than migraine.

Neurogenic inflammation in the dura mater during migraine

Comparable findings in animals and man related to the release of neuropeptides and the action of antimigraine medications, such as ergots and triptans, provide support for the validity of the animal studies of neurogenic inflammation. However, a remaining key issue is the actual demonstration of plasma-protein extravasation during migraine headache. Plasma-protein extravasation occurs in the retina of the rat after trigeminal stimulation, a response that can be significantly reduced by pre-treatment with sumatriptan [72]. Since it is easily accessible in man by angiography, plasma-protein extravasation was investigated in the retina during migraine and cluster headache but with negative results. Although interesting, this particular observation may, however, not be relevant to the pathogenesis of the migraine headache because the eye receives nerve fibers from a different population of trigeminal neurons than the dura mater [73]. The finding that sumatriptan reduces retinal edema in the rat merely reflects the presence of serotonin-1B and -D receptors on the trigeminal nerve fibers that innervate the eye.

Very preliminary but nevertheless encouraging evidence for the occurrence of plasma-protein extravasation in man was recently reported using single-photon emission computed tomography with 99-technetium labeled albumin [74]. It concerns an imaging technique that measures signal intensity provided by the accumulation of radioactive tracer. The two migraineurs in the study were injected with albumin labeled with 99-technetium at the beginning of an attack and images were obtained immediately and 4 h later. While no signal could be obtained with the first scan, the scan at 4 h revealed accumulation of tracer in intracranial extraparenchymal tissue (dura mater?), ipsilateral to the headache. The presence of extravasated albumin during migraine headache ipsilateral to the pain suggests that neurogenic inflammation does, indeed, play a role in the pathogenesis of the migraine headache. In addition, Goebel et al. recently reported plasma-protein extravasation around the cavernous sinus in 16 patients with cluster headache, during headache and ipsilateral to the pain, using the same technique [75].

Electrophysiological properties of the trigeminal neurons

The dura mater is densely innervated by small unmyelinated C-fibers and a model has been developed to assess the electrophysiological properties of the neurons involved. Strassman et al. studied the properties of the neurons of the trigeminal ganglion and Burstein et al. investigated those contained in the trigeminal nucleus caudalis [76, 77]. Electrical or mechanical stimulation of exposed dura mater or dural blood vessels in the rat or cat identified the projecting neurons with cell bodies in the trigeminal ganglion or trigeminal nucleus caudalis. In order to carry out the brainstem experiments, the bone around the transverse and sagittal sinuses was removed, the upper cervical spinal cord exposed, and the dura as well as pia mater above the spinal cord removed. Once a dura-responsive neuron in the trigeminal nucleus caudalis had been identified that also had a receptive field in the face, the properties of the neuron were recorded. Sensory modalities tested were the Von Frey hair applied to the dura mater and pain, heat, and brush stimulation of the face.

In the experiments, several substances were identified that sensitize trigeminal neurons, among them acetic phosphate buffer (pH 4.7) and a mixture of histamine, serotonin and bradykinin (10^{-3} mol) that also contained prostaglandin E2 (10^{-4} mol) (pH 5), referred to as "inflammatory soup". After brief exposure of the dura mater to the acetic phosphate buffer (2 min), the inflammatory soup was applied for 2 min. Sensitization developed within 20–30 min and lasted up to 90 min after the soup had been removed, consisting of enhanced sensitivity of the dura mater and face to mechanical stimulation with expansion of receptive fields. In addition, neurons unresponsive to mechanical stimulation before sensitization became sensitive to this kind of stimulation after application of the inflammatory soup [76]. It is interesting that application of lidocaine to the dura mater reduced the mechanical sensitivity of the dura but did not affect the sensitization to mechanical or thermal stimuli of the face, suggesting a central mechanism in maintaining the sensitized state of the neurons involved.

The above findings certainly explain a commonly observed phenomenon during migraine headache. Non-painful action of blood vessels, such as the pulsation associated with the heart beat, could be perceived as painful due to enhanced sensitivity of the neurons to mechanical stimulation, which could be caused by inflammation of the dura [77]. Thus, inflammation of the dura mater can result in activation of central neurons, which, in turn, can lead to extracranial allodynia referring to non-painful stimuli becoming painful. This phenomenon was also investigated during migraine headache using mechanical and thermal stimulation; in about 80% of the migraineurs studied, cutaneous allodynia was found in the ipsilateral side of the head [78].

The above findings provide evidence that the pathogenesis of the migraine headache involves not only irritation of perivascular nociceptive fibers in the dura mater but also a transient increase in responsiveness, or sensitization, of central neurons involved in processing pain arising from intra- and extracranial structures. The interpretation of the experiments was based, however, on the assumption that there is inflammation in the dura mater during migraine headache, which still needs to be conclusively demonstrated.

Prior to the detection that neurons responsive to stimulation of the dura mater become sensitized, stimulation of the dura simply identified responsive trigeminal neurons. In 1986, electrical stimulation of the dura led to the identification of responsive trigeminal neurons in the trigeminal nucleus caudalis [79]. Activation of extracellular recording in the trigeminal nucleus caudalis in the cat, after stimulation of the dura mater, has also resulted in the identification of neurons with orofacial receptive fields [80].

Stimulation of the superior sagittal sinus

Geppetti et al. showed that administration of capsaicin caused *in vitro* release of neuropeptides into superfusates of dural venous sinuses in the guinea pig [15]. Stimulation of the superior sagittal sinus also caused elevations in neuropeptide levels in blood drawn from the external jugular vein [45], which was also observed during migraine headache [49]. Based on the findings by Ray and Wolff, who in man observed dense trigeminal innervation of the superior sagittal sinus [81], it was concluded that neuropeptides might also be released here from trigeminal nerve fibers.

Electrical or mechanical stimulation of the superior sagittal sinus caused activation of the dorsolateral part of the C2 spinal segment in the cat [82, 83]. Lidocaine injections into or lesioning of the trigeminal ganglion abolished the response to mechanical stimulation. Antimigraine medications, such as sumatriptan – but after disruption of the blood-brain barrier only – zolmitriptan and, to a lesser extent, aspirin reduced the peak-to-peak amplitude of evoked potentials after stimulation of the superior sagittal sinus in the cat, suggesting a central mode of action [84–86].

Stimulation of the superior sagittal sinus also caused c-fos expression in the trigeminal nucleus caudalis [82] and a role of nitric oxide in this effect was discovered in 1999. Pre-treatment of the cat with the inhibitor of nitric oxide synthase, n(g)-nitro-l-arginine, 30 min before electrical stimulation of the sagittal sinus, markedly reduced the number of cells positive for c-fos in the trigeminal nucleus caudalis [87]. This suggests a mediator effect of nitric oxide in the activation of the nociceptive trigeminal neurons.

Ergots and sumatriptan also reduced the expression of c-fos in the trigeminal nucleus caudalis after stimulation of the superior sagittal sinus. Surprisingly, a lipophilic neurokinin-1-receptor antagonist, GR-205171, in a dose of 100 µg/kg intravenously, did not affect the c-fos expression after stimulation of the superior sagittal sinus or the rate of impulse transmission within trigeminal nucleus caudalis of the cat. However, GR-205171 (100 µg/kg intravenously) reduced c-fos expression after trigeminal stimulation and administration of capsaicin in the guinea pig [88]. Furthermore, GR-205171 dose-dependently reduced plasma-protein extravasation after trigeminal stimulation in the rat (0.1–1 mg/kg intravenously) and the guinea pig (1–100 µg/kg intravenously) [89].

Cortical blood flow

The specific antimigraine medications, ergots and triptans, are potent vasoconstrictors and may exert their beneficial effect through constricting cranial arteries and possibly also arteriovenous anastomoses [90]. In accordance with this, the pain of the migraine headache has been attributed to dilation of cranial arteries or opening of arteriovenous anastomoses [81, 92]. At least in the pig, the triptans decrease total carotid blood flow almost exclusively at the expense of flow through arteriovenous anastomoses [93].

Jansen et al. examined the effects of calcium-entry blockers and Edvinsson focused on those of neuropeptides on the isolated human cerebral artery [94, 95]. Alpha-adrenergic and serotonergic mechanisms were investigated in isolated human temporal arteries and no differences were found between migraineurs and controls [96].

Stimulation of the trigeminal ganglion increases cortical blood flow in the brain, presumably due to peptide-mediated vasodilation. The increase in cortical blood flow induced by electrical stimulation of the nasociliary nerve, a branch of the trigeminal nerve, is frequency-dependent and markedly inhibited, in two out of three studies, by the receptor antagonist for calcitonin gene-related peptide, calcitonin gene-related peptide 8-37 [97–99]. Williamson et al. used intravital microscopy to measure the diameter of dural blood vessels after stimulating the surface of the dura mater exposed through a cranial window. The vasodilation was blocked by superfusion with the receptor antagonist for calcitonin gene-related peptide, calcitonin gene-related peptide 8-37 [13]. Thus, it is likely that vasodilation in the dura mater is caused by activation of trigeminal nerve fibers and, at least in part, mediated by the release of calcitonin gene-related peptide.

Kurosawa et al. [100] and McCulloch [101] came to the same conclusion and their findings are consistent with the results of meningeal activation using *Streptococcus pneumoniae* or injury induced by acute, severe hypertension or seizures [13, 102, 103]. In the latter experiments, trigeminal nerve fibers were stimulated by injury either to blood vessels or brain tissue, resulting in the release of neuropeptides from the nerve fibers and an increase in cortical blood flow. Sectioning of the trigeminal nerve but not that of the trigeminal nerve root abolished the increase in cortical blood flow. Section of the nerve preserves cell bodies and perivascular nerve fibers, whereas these structures die after sectioning of the nerve root.

Cortical spreading depression

Cortical spreading depression is thought to be the pathophysiological correlate of the symptoms of the migraine aura. It is a slowly propagating wave of transient suppression of cortical neuronal activity, with loss of membrane potential and accompanied by a major metabolic disturbance [104]. The electrophysiological changes due to cortical spreading depression are associated with short-lasting hyperperfusion, followed by prolonged hypoperfusion.

Lashley, studying his own visual migraine aura, observed a spread of scintillations from the center to the periphery of his visual field at a speed of approximately 3 mm per minute, followed by a temporary decrease of vision. The similarities between the visual aura symptoms and the features of cortical spreading depression made Milner suggest that cortical spreading depression may be the underlying mechanism of the migraine aura [104]. Since then, cortical spreading depression has been considered a useful phenomenon to investigate in animals in an attempt to understand the pathogenesis of the migraine aura.

Cortical spreading depression was first described by Leão in the rabbit in 1944 [105] and has since been studied extensively, mostly in the cat. Recently, cortical spreading depression in the cat was investigated using diffusion-weighted echo-planar magnetic resonance imaging, revealing primary and secondary events related to it with different spatial extent, velocity and spread [106]. Cortical spreading depression can be elicited by the application of chemicals, for example, potassium, as well as by electrical or mechanical stimulation of the cerebral cortex (for review, see [107]). A phenomenon like cortical spreading depression, peri-infarct depolarization, has been shown to contribute to tissue damage in stroke [91, 108], although cortical spreading depression in the normal brain does not cause tissue injury [107].

In the last decade, cortical spreading depression has also been shown to occur in man under pathological circumstances. In 1991, Avoli reported cortical spreading depression in human brain slices *in vitro* [109] and Mayevski, 5 years later, observed the phenomenon in patients following head injury [110].

Recent studies with magnetic resonance imaging during visually triggered migraine headaches showed bilateral suppression of neuronal activity and increased energy demand before the onset of headache, slowly traveling across the occipital cortex [111]. However, the most compelling evidence for the existence of cortical spreading depression as the underlying mechanism of the migraine aura was reported by Hadjikhani et al. [112]. In their study, the visual migraine aura following moderate exercise was explored using functional magnetic resonance imaging. By using a visual stimulation-dependent paradigm, the blood oxygenation-level dependent (BOLD) signal showed an initial increase in the mean signal and a decrease in amplitude. The most likely source of this initial response was an increase in blood flow and blood volume due to heightened neuronal activity, lasting for 3–4.5 minutes. The change in signal occurred exclusively in the occipital lobe contralateral to the visual aura. It started posteriorly and moved forward following the progression of the aura at a rate calculated to be 3.5 mm per minute, consistent with the spread of cortical spreading depression in animals. The initial rise of the signal, reflecting the wave of excitation accompanied by hyperemia, implies that the aura is caused by an increase in neuronal activity and is not due to ischemia.

In migraine, the aura is typically followed by headache, but it is not clear if cortical spreading depression can activate the trigeminal system and, hence, can lead to pain. Cortical spreading depression causes c-fos expression in the trigeminal nucleus caudalis, which was markedly reduced after transsection of the trigeminal nerve or pre-treatment with sumatriptan [113]. In addition, it was observed that the hyperperfusion associated with cortical spreading depression is mediated by the release of calcitonin gene-related peptide, suggesting activation of trigeminal nerve fibers [114]. However, Ebersberger et al., using single-unit recording in the rat, failed to show activation of neurons in the trigeminal nucleus caudalis after potassium-induced cortical spreading depression [115].

Familial hemiplegic migraine and episodic ataxia type 2 have been linked to mutations of the α_1 subunit of the voltage-dependent calcium channel (CACNL1A4) [116]. The mutation underlying the phenotype of the tottering as well as of the leaner mouse also results in a defect in the same calcium-channel α_1 subunit [117]. Potassium-evoked release of the neurotransmitters, glutamate and GABA, are substantially reduced in both mutant mice [118]. In addition, the mice are resistant to the induction of cortical spreading depression, a phenomenon involving, among others, the release of the excitatory neurotransmitter, glutamate, from the cortical neuron [119].

Conclusion

We described most of the currently used animal models to investigate the pathogenesis of migraine, each developed to answer specific questions. While electrical and chemical stimulation of the trigeminal system are methods primarily accepted to investigate mechanisms related to neurogenic inflammation, other models, such as that of stimulation of the superior sagittal sinus, focus on the electrophysiological properties of the secondary trigeminal neurons in the brainstem. Each of the models has contributed to our understanding of the pathogenesis of migraine and numerous features detected in animals have also been observed in man during migraine headache. In particular, neuropeptide release has been shown in several animal models, as well as in man during migraine headache and these peptides are essential for the development of dural inflammation. Moreover, powerful medications that abort migraine headache have also been shown to attenuate neuropeptide release and, subsequently, have been shown to inhibit neurogenic inflammation in animals. These findings in particular strengthen the notion that migraine is a neuroinflammatory disease.

References

1 Mayberg M, Langer RS, Zervas NT, Moskowitz MA (1981) Perivascular meningeal projections from cat trigeminal ganglia: possible pathway for vascular headaches in man. *Science* 213: 228–230

2 Feindel W, Penfield W, McNaughton F (1960) The tentorial nerves and localization of intracranial pain in man. *Neurology* 10: 555–563

3 Mayberg MR, Zervas NT, Moskowitz MA (1984) Trigeminal projections to supratentorial pial and dural blood vessels in cats demonstrated by horseradish peroxidase histochemistry. *J Comp Neurol* 223: 46–56

4 Uddman R, Edvinsson L (1989) Neuropeptides in the cerebral circulation. *Cerebrovasc Brain Metab Rev* 1: 230–252

5 Liu-Chen LY, Liszczak TM, King JC, Moskowitz MA (1986) Immunoelectron microscopic study of substance P-containing fibers in feline cerebral arteries. *Brain Res* 369: 12–20

6 Jansen I, Alafaci C, McCulloch J et al (1991) Tachykinins (substance P, neurokinin A,

neuropeptide K, and neurokinin B) in the cerebral circulation: vasomotor responses *in vitro* and *in situ*. *J Cereb Blood Flow Metab* 11: 567–575
7 Dimitriadou V, Buzzi MG, Moskowitz MA, Theoharides TC (1991) Trigeminal sensory fiber stimulation induces morphological changes reflecting secretion in rat dura mater mast cells. *Neuroscience* 44: 97–112
8 Dimitriadou V, Buzzi MG, Theoharides TC, Moskowitz MA (1992) Ultrastructural evidence for neurogenically mediated changes in blood vessels of the rat dura mater and tongue following antidromic trigeminal stimulation. *Neuroscience* 48: 187–203
9 Markowitz S, Saito K, Moskowitz MA (1987) Neurogenically mediated leakage of plasma protein occurs from blood vessels in dura mater but not brain. *J Neurosci* 7: 4129–4136
10 Yu XJ, Waeber C, Castanon N et al (1996) 5-Carboxamido-tryptamine, CP-122,288 and dihydroergotamine but not sumatriptan, CP-93,129, and serotonin-5-O-carboxymethyl-glycyl-tyrosinamide block dural plasma protein extravasation in knockout mice that lack 5-hydroxytryptamine$_{1B}$ receptors. *Mol Pharmacol* 49: 761–765
11 Johnson KW, Phebus LA (1998) A fluorescence-based method for assessing dural protein extravasation induced by trigeminal ganglion stimulation. *J Neurosci Methods* 81: 19–24
12 Huang Z, Byun B, Matsubara T, Moskowitz MA (1993) Time-dependent blockade of neurogenic plasma extravasation in dura mater by 5-HT$_{1B/D}$ agonists and endopeptidase. *Br J Pharmacol* 108: 331–335
13 Williamson DJ, Hargreaves RJ, Hill RG, Shepheard SL (1997) Intravital microscope studies on the effects of neurokinin agonists and calcitonin gene-related peptide on dural vessel diameter in the anaesthetized rat. *Cephalalgia* 17: 518–524
14 Buzzi MG, Dimitriadou V, Theoharides TC, Moskowitz MA (1992) 5-Hydroxytryptamine receptor agonists for the abortive treatment of vascular headaches block mast cell, endothelial and platelet activation within the rat dura mater after trigeminal stimulation. *Brain Res* 583: 137–149
15 Geppetti P, Del Bianco E, Santicioli P et al (1990) Release of sensory neuropeptides from dural venous sinuses of guinea pig. *Brain Res* 510: 58–62
16 Russell LC, Burchiel KJ (1984) Neurophysiological effects of capsaicin. *Brain Res* 320: 165–176
17 Moskowitz MA, Brody M, Liu-Chen LY (1983) *In vitro* release of immunoreactive substance P from putative afferent nerve endings in bovine pia arachnoid. *Neuroscience* 9: 809–814
18 Buzzi MG, Moskowitz MA (1990) The antimigraine drug, sumatriptan (GR43175), selectively blocks neurogenic plasma extravasation from blood vessels in dura mater. *Br J Pharmacol* 99: 202–206
19 Markowitz S, Saito K, Moskowitz MA (1988) Neurogenically mediated plasma extravasation in dura mater: effect of ergot alkaloids. A possible mechanism of action in vascular headache. *Cephalalgia* 8: 83–91
20 Gamse R, Holzer P, Lembeck F (1980) Decrease of substance P in primary afferent neu-

rons and impairment of neurogenic plasma extravasation by capsaicin. *Br J Pharmacol* 68: 207–213
21 Markowitz S, Saito K, Buzzi MG, Moskowitz MA (1989) The development of neurogenic plasma extravasation in the rat dura mater does not depend upon the degranulation of mast cells. *Brain Res* 477: 157–165
22 Williams S, Evan G, Hunt SP (1990) Spinal c-fos induction by sensory stimulation in neonatal rats. *Neurosci Lett* 109: 309–314
23 Nozaki K, Moskowitz MA, Boccalini P (1992) CP-93,129, sumatriptan, dihydroergotamine block c-fos expression within rat trigeminal nucleus caudalis caused by chemical stimulation of the meninges. *Br J Pharmacol* 106: 409–415
24 Cutrer FM, Moussaoui S, Garret C, Moskowitz MA (1995) The non-peptide neurokinin-1 antagonist, RPR 100893, decreases c-fos expression in trigeminal nucleus caudalis following noxious chemical meningeal stimulation. *Neuroscience* 64: 741–750
25 Cutrer FM, Schoenfeld D, Limmroth V et al (1995) Suppression by the sumatriptan analogue, CP-122,288 of c-fos immunoreactivity in trigeminal nucleus caudalis induced by intracisternal capsaicin. *Br J Pharmacol* 114: 987–992
26 Tassorelli C, Joseph SA (1995) Systemic nitroglycerin induces Fos immunoreactivity in brainstem and forebrain structures of the rat. *Brain Res* 682: 167–181
27 Tassorelli C, Joseph SA (1995) NADPH-diaphorase activity and fos expression in brain nuclei following nitroglycerin administration. *Brain Res* 695: 37–44
28 Olesen J, Iversen HK, Thomsen LL (1993) Nitric oxide supersensitivity: a possible molecular mechanism of migraine pain. *Neuroreport* 4: 1027–1030
29 Lassen LH, Ashina M, Christiansen I et al (1997) Nitric oxide synthase inhibition in migraine [letter]. *Lancet* 349: 401–402
30 Knyihar-Csillik E, Vecsei L (1999) Effect of a nitric oxide donor on nitroxergic nerve fibers in the rat dura mater. *Neurosci Lett* 260: 97–100
31 Buzzi MG, Carter WB, Shimizu T et al (1991) Dihydroergotamine and sumatriptan attenuate levels of CGRP in plasma in rat superior sagittal sinus during electrical stimulation of the trigeminal ganglion. *Neuropharmacology* 30: 1193–1200
32 Saito K, Markowitz S, Moskowitz MA (1988) Ergot alkaloids block neurogenic extravasation in dura mater: proposed action in vascular headaches. *Ann Neurol* 24: 732–737
33 Buzzi MG, Moskowitz MA (1991) Evidence for 5-$HT_{1B/1D}$ receptors mediating the antimigraine effect of sumatriptan and dihydroergotamine. *Cephalalgia* 11: 165–168
34 Nilsson T, Longmore J, Shaw D et al (1999) Contractile 5-HT_{1B} receptors in human cerebral arteries: pharmacological characterization and localization with immunocytochemistry. *Br J Pharmacol* 128: 1133–1140
35 Razzaque Z, Heald MA, Pickard JD et al (1999) Vasoconstriction in human isolated middlemeningeal arteries: determining the contribution of 5-HT_{1B}- and 5-HT_{1F}-receptor activation. *Br J Clin Pharmacol* 47: 75–82
36 Shepheard S, Edvinsson L, Cumberbatch M et al (1999) Possible antimigraine mechanisms of action of the $5HT_{1F}$ receptor agonist LY334370. *Cephalalgia* 19: 851–858

37 Johnson KW, Schaus JM, Durkin MM et al (1997) 5-HT_{1F} receptor agonists inhibit neurogenic dural inflammation in guinea pigs. *Neuroreport* 8: 2237–2240
38 Goldstein DJ, Wang O, Saper JR et al (1997) Ineffectiveness of neurokinin-1 antagonist in acute migraine: a crossover study. *Cephalalgia* 17: 785–790
39 Martin GR, Robertson AD, MacLennan SJ et al (1997) Receptor specificity and trigemino-vascular inhibitory actions of a novel 5-$HT_{1B/1D}$ receptor partial agonist, 311C90 (zolmitriptan). *Br J Pharmacol* 121: 157–164
40 Williamson DJ, Shepheard SL, Hill RG, Hargreaves RJ (1997) The novel anti-migraine agent rizatriptan inhibits neurogenic dural vasodilation and extravasation. *Eur J Pharmacol* 328: 61–64
41 Buzzi MG, Sakas DE, Moskowitz MA (1989) Indomethacin and acetylsalicylic acid block neurogenic plasma protein extravasation in rat dura mater. *Eur J Pharmacol* 165: 251–258
42 Lobo BL, Landy S (1994) Recommendations for the emergency treatment of migraine headache. *J Tenn Med Assoc* 87: 53–54
43 Gallagher RM (1986) Emergency treatment of intractable migraine. *Headache* 26: 74–75
44 Mason RT, Peterfreund RA, Sawchenko PE et al (1984) Release of the predicted calcitonin gene-related peptide from cultured rat trigeminal ganglion cells. *Nature* 308: 653–655
45 Goadsby PJ, Edvinsson L, Ekman R (1988) Release of vasoactive peptides in the extracerebral circulation of humans and the cat during activation of the trigeminovascular system. *Ann Neurol* 23: 193–196
46 Zagami AS, Goadsby PJ, Edvinsson L (1990) Stimulation of the superior sagittal sinus in the cat causes release of vasoactive peptides. *Neuropeptides* 16: 69–75
47 Ebersberger A, Averbeck B, Messlinger K, Reeh PW (1999) Release of substance P, calcitonin gene-related peptide and prostaglandin E2 from rat dura mater encephali following electrical and chemical stimulation *in vitro*. *Neuroscience* 89: 901–907
48 Durham PL, Russo AF (1999) Regulation of calcitonin gene-related peptide secretion by a serotonergic antimigraine drug. *J Neurosci* 19: 3423–3429
49 Goadsby PJ, Edvinsson L, Ekman R (1990) Vasoactive peptide release in the extracerebral circulation of humans during migraine headache. *Ann Neurol* 28: 183–187
50 Goadsby PJ, Edvinsson L (1993) The trigeminovascular system and migraine: studies characterizing cerebrovascular and neuropeptide changes seen in humans and cats. *Ann Neurol* 33: 48–56
51 Knyihar-Csillik E, Tajti J, Samsam M et al (1998) Depletion of calcitonin gene-related peptide from the caudal trigeminal nucleus of the rat after electrical stimulation of the Gasserian ganglion. *Exp Brain Res* 118: 111–114
52 Samsam M, Covenas R, Ahangari R et al (1999) Alterations in neurokinin A-, substance P- and calcitonin gene-related peptide immunoreactivities in the caudal trigeminal nucleus of the rat following electrical stimulation of the trigeminal ganglion. *Neurosci Lett* 261: 179–182

53 Pedersen-Bjergaard U, Nielsen LB, Jensen K et al (1989) Algesia and local responses induced by neurokinin A and substance P in human skin and temporal muscle. *Peptides* 10: 1147–1152

54 Pedersen-Bjergaard U, Nielsen LB, Jensen K et al (1991) Calcitonin gene-related peptide, neurokinin A and substance P: effects on nociception and neurogenic inflammation in human skin and temporal muscle. *Peptides* 12: 333–337

55 McDuffie JE, Coaxum SD, Maleque MA (1999) 5-hydroxytryptamine evokes endothelial nitric oxide synthase activation in bovine aortic endothelial cell cultures. *Proc Soc Exp Biol Med* 221: 386–390

56 Ishida T, Kawashima S, Hirata K, Yokoyama M (1998) Nitric oxide is produced *via* 5-HT_{1B} and 5-HT_{2B} receptor activation in human coronary artery endothelial cells. *Kobe J Med Sci* 44: 51–63

57 Stepien A, Chalimoniuk M, Strosznajder J (1999) Serotonin $5HT_{1B/1D}$ receptor agonists abolish NMDA receptor-evoked enhancement of nitric oxide synthase activity and cGMP concentration in brain cortex slices. *Cephalalgia* 19: 859–865

58 Wei EP, Moskowitz MA, Boccalini P, Kontos HA (1992) Calcitonin gene-related peptide mediates nitroglycerin and sodium nitroprusside-induced vasodilation in feline cerebral arterioles. *Circ Res* 70: 1313–1319

59 Lee WS, Limmroth V, Ayata C et al (1995) Peripheral GABA receptor-mediated effects of sodium valproate on dural plasma protein extravasation to substance P and trigeminal stimulation. *Br J Pharmacol* 116: 1661–1667

60 Limmroth V, Lee WS, Moskowitz MA (1996) GABAA-receptor-mediated effects of progesterone, its ring-A-reduced metabolites and synthetic neuroactive steroids on neurogenic oedema in the rat meninges. *Br J Pharmacol* 117: 99–104

61 Okuno T, Itakura T, Lee TJ et al (1994) Cerebral pial arterial innervation with special reference to GABAergic innervation. *J Auton Nerv Syst* 49 (Suppl): S105–S110

62 Cutrer FM, Moskowitz MA (1996) The actions of valproate and neurosteroids in a model of trigeminal pain. *Headache* 36: 579–585

63 Bileviciute I, Stenfors C, Theodorsson E, Lundeberg T (1998) Unilateral injection of calcitonin gene-related peptide (CGRP) induces bilateral oedema formation and release of CGRP-like immunoreactivity in the rat hindpaw. *Br J Pharmacol* 125: 1304–1312

64 Bourgoin S, Pohl M, Benoliel JJ et al (1992) Gamma-aminobutyric acid, through $GABA_A$ receptors, inhibits the potassium-stimulated release of calcitonin gene-related pep. *Brain Res* 583: 344–348

65 Akasu T, Munakata Y, Tsurusaki M, Hasuo H (1999) Role of $GABA_A$ and $GABA_C$ receptors in the biphasic GABA responses in neurons of the rat major pelvic ganglia. *J Neurophysiol* 82: 1489–1496

66 Kataoka Y, Niwa M, Yamashita K, Taniyama K (1994) GABA receptor function in the parasympathetic ganglia. *Jpn J Physiol* 44 (Suppl 2): S125–S129

67 Kawatani M, Whitney T, Booth AM, De Groot WC (1989) Excitatory effect of substance P in parasympathetic ganglia of cat urinary bladder. *Am J Physiol* 257: R1450–R1456

100 Kurosawa M, Messlinger K, Pawlak M, Schmidt RF (1995) Increase of meningeal blood flow after electrical stimulation of rat dura mater encephali: mediation by calcitonin gene-related peptide. *Br J Pharmacol* 114: 1397–1402
101 McCulloch J, Uddman R, Kingman TA, Edvinsson L (1986) Calcitonin gene-related peptide: functional role in cerebrovascular regulation. *Proc Natl Acad Sci USA* 83: 5731–5735
102 Sakas DE, Moskowitz MA, Wei EP et al (1989) Trigeminovascular fibers increase blood flow in cortical gray matter by axon reflex-like mechanisms during acute severe hypertension or seizures. *Proc Natl Acad Sci USA* 86: 1401–1405
103 Macfarlane R, Tasdemiroglu E, Moskowitz MA et al (1991) Chronic trigeminal ganglionectomy or topical capsaicin application to pial vessels attenuates postocclusive cortical hyperemia but does not influence postischemic hypoperfusion. *J Cereb Blood Flow Metab* 11: 261–271
104 Moskowitz MA (1984) The neurobiology of vascular head pain. *Ann Neurol* 16: 157–168
105 Leão AAP (1944) Spreading depression of activity in the cerebral cortex. *J Neurophysiol* 7: 359–390
106 Bockhorst KH, Smith JM, Smith MI et al (2000) A quantitative analysis of cortical spreading depression events in the feline brain characterized with diffusion-weighted MRI. *J Magn Reson Imaging* 12: 722–733
107 Lauritzen M (1987) Cerebral blood flow in migraine and cortical spreading depression. *Acta Neurol Scand* 113: 1–40
108 Mies G, Iijima T, Hossmann KA (1993) Correlation between peri-infarct DC shifts and ischaemic neuronal damage in rat. *Neuroreport* 4: 709–711
109 Avoli M, Drapeau C, Louvel J et al (1991) Epileptiform activity induced by low extracellular magnesium in the human cortex maintained *in vitro*. *Ann Neurol* 30: 589–596
110 Mayevsky A, Doron A, Manor T et al (1996) Cortical spreading depression recorded from the human brain using a multiparametric monitoring system. *Brain Res* 740: 268–274
111 Cao Y, Welch KM, Aurora S, Vikingstad EM (1999) Functional MRI-BOLD of visually triggered headache in patients with migraine. *Arch Neurol* 56: 548–554
112 Hadjikhani N, Sanchez del Rio M, Wu O, Schwartz D, Bakker D, Fischl B, Kwong KK, Cutrer FM, Rosen BR, Tootell RB et al (2001) Mechanisms of migraine aura revealed by functional MRI in human visual cortex. *Proc Natl Acad Sci USA* 98 (8): 4687–4692
113 Moskowitz MA, Nozaki K, Kraig RP (1993) Neocortical spreading depression provokes the expression of c-fos protein-like immunoreactivity within trigeminal nucleus caudalis *via* trigeminovascular mechanisms. *J Neurosci* 13: 1167–1177
114 Reuter U, Weber JR, Gold L et al (1998) Perivascular nerves contribute to cortical spreading depression-associated hyperemia in rats. *Am J Physiol* 274: H1979–H1987
115 Ebersberger A, Richter F, Averbeck B, Schaible H-G (2000) The trigeminal vascular system is not activated nor sensitized by cortical spreading depression. *Soc Neurosci* 26: 355–356

116 Ophoff RA, Terwindt GM, Vergouwe MN et al (1996) Familial hemiplegic migraine and episodic ataxia type-2 are caused by mutations in the Ca^{2+} channel gene CACNL1A4. *Cell* 87: 543–552
117 Fletcher CF, Lutz CM, O'Sullivan TN et al (1996) Absence epilepsy in tottering mutant mice is associated with calcium channel defects. *Cell* 87: 607–617
118 Shimizu-Sasamata M, Lo HE, Noebels JL, Moskowitz MA (1997) P/Q-type calcium channel mutation reduces potassium evoked cortical neurotransmitter release. *Soc Neurosci* 27: 1182–1182
119 Ayata C, Shimizu-Sasamata M, Lo EH et al (2000) Impaired neurotransmitter release and elevated threshold for cortical spreading depression in mice with mutations in the alpha 1A subunit of P/Q type calcium channels. *Neuroscience* 95: 639–645
120 Goadsby PJ, Duckworth JW (1987) Effect of stimulation of trigeminal ganglion on regional cerebral blood flow in cats. *Am J Physiol* 253: R270–R274
121 Uhl GR, Walther D, Nishimori T et al (1991) Jun B, c-jun, jun D and c-fos mRNAs in nucleus caudalis neurons: rapid selective enhancement by afferent stimulation. *Brain Res Mol Brain Res* 11: 133–141
122 Matsubara T, Moskowitz MA, Huang Z (1992) UK-14,304, R(-)-alpha-methyl-histamine and SMS 201-995 block plasma protein leakage within dura mater by prejunctional mechanisms. *Eur J Pharmacol* 224: 145–150
123 Shimizu-Sasamata M, Bosque-Hamilton P, Huang PL et al (1998) Attenuated neurotransmitter release and spreading depression-like depolarizations after focal ischemia in mutant mice with disrupted type I nitric oxide synthase gene. *J Neurosci* 18: 9564–9571
124 Skarby T, Tfelt-Hansen P, Gjerris F, et al (1982) Characterization of 5-hydroxytryptamine receptors in human temporal arteries: comparison between migraine sufferers and nonsufferers. *Ann Neurol* 12: 272–277

Mediators and their receptors involved in neurogenic inflammation

Dimos D. Mitsikostas

Athens Naval Hospital, Headache Clinic, 70 Dinokratous Street, 11521 Athens, Greece

Introduction

Trigeminal system

The trigeminal system is the basic neuronal system that transmits nociceptive information from the cranial structures, *via* the trigeminal ganglion, to the trigeminal nucleus caudalis [1]. Based on their histochemical and functional properties, the trigeminal ganglion cells are divided into two groups: approximately one-third project large, myelinated A-fibers and are mostly sensitive to mechanical stimulation; the remaining two-thirds project small, unmyelinated Aδ- and C-fibers and are nociceptive. The nerve cells projecting the small fibers can be further divided into two groups of approximately equal size. One subgroup contains calcitonin gene-related peptide and substance P as well as their receptors and also receptors for nerve growth factor. The remaining cells have a distinct chemical phenotype, are non-peptidergic, and carry receptors for the neurotrophic factor derived from the glial cell line [2].

The peripheral endings of the Aδ- and C-fibers surround the dural blood vessels and together with them constitute a nociceptive, neurovascular network known as the trigeminovascular system [1, 3]. When activated, the nerve fibers release vasoactive neuropeptides from the perivascular nerve endings, calcitonin gene-related peptide, neurokinin A and substance P, synthesized by messenger RNA and ribosomal mechanisms in the cell bodies of the trigeminal ganglion. The neuropeptides are released by a calcium-dependent mechanism, resulting in so-called neurogenic inflammation of the dural blood vessels, consisting of vasodilation, plasma-protein extravasation, degranulation of mast cells, and platelet aggregation [4–7].

Neurogenic inflammation

Experimentally, dural neurogenic inflammation can be evoked by antidromic electrical, mechanical, or chemical stimulation of the sensory nerve fibers [6]. Electrical stimulation of the trigeminal ganglion results in plasma-protein extravasation in the dura mater [5], quantified on the basis of leakage of albumin, labeled with radioactive trac-

Migraine: A Neuroinflammatory Disease?, edited by Egilius L.H. Spierings and Margarita Sánchez del Río

er, from the dural blood vessels. Stimulation of the sphenopalatine ganglion also results in dural neurogenic inflammation *via* muscarinic cholinergic receptors, suggesting that activation of the parasympathetic nervous system is involved as well [8].

The plasma-protein extravasation is attenuated by several medications clinically effective in migraine, such as the non-steroidal anti-inflammatory analgesics, indomethacin and aspirin [9], the ergots, dihydroergotamine and ergotamine [10, 11], the triptans, sumatriptan [11], zolmitriptan [12], naratriptan [13] and rizatriptan [14], and the anticonvulsant, valproate [15]. Thus, the model of dural neurogenic inflammation could serve as an experimental model of migraine, although there are now a number of compounds that are effective in inhibiting plasma-protein extravasation but are not effective in aborting migraine headache. The purpose of this chapter is to review the mediators known to be involved in dural neurogenic inflammation in animals, focusing on the role of calcitonin gene-related peptide, substance P, serotonin, nerve growth factor, and γ-aminobutyric acid (GABA), together with their relevant receptor systems.

Calcitonin gene-related peptide

Effects

Calcitonin gene-related peptide is an inflammatory and vasoactive, 37-amino-acid peptide, the most potent vasodilator among the neuropeptides [16], localized in the cerebral vasculature [17] and neuronal tissues, constituting the trigeminal ganglion [18–20] and trigeminal nucleus caudalis [21]. Apart from its vascular effects, calcitonin gene-related peptide may also be involved in nociception, glucose re-uptake, and stimulation of glycolysis in skeletal muscle [22]. Like other neuropeptides, it exerts its effects by binding to specific membrane receptors, namely the calcitonin gene-related peptide 1 and 2 receptor [23], located in the vascular endothelium and trigeminal ganglion [23, 24]. There is evidence that nerve growth factor modulates the synthesis of calcitonin gene-related peptide and that release of the peptide is under serotonergic control. At the level of the spinal cord, the proteinase-activated type 2 receptor also modulates the release of calcitonin gene-related peptide, indicating that proteases may induce dural neurogenic inflammation *via* the release of calcitonin gene-related peptide [25].

Synergism

Most neurons in the sensory ganglia immunoreactive to substance P also contain calcitonin gene-related peptide; the two peptides are released together and act synergistically:

1. Calcitonin gene-related peptide increases the plasma-protein extravasation induced by substance P in the rat skin;
2. Behaviorally, intrathecally administered calcitonin gene-related peptide potentiates scratching and biting behavior induced by substance P and both peptides modulate the nociceptive flexor reflex; and
3. Calcitonin gene-related peptide inhibits the enzyme that hydrolyzes substance P and potentiates the release of this peptide as well as that of glutamate in the dorsal horn [26].

Release

Electrical stimulation of the trigeminal ganglion causes the release of calcitonin gene-related peptide in the external jugular vein in animals and man [27]. In cats, increased levels of calcitonin gene-related peptide in the external jugular vein have been reported after stimulation of the superior sagittal sinus [28]. In man, calcitonin gene-related peptide levels were elevated during migraine headache in the external jugular vein, returning to baseline after resolution of the pain [29]. The plasma level of calcitonin gene-related peptide has also been reported to be elevated between migraine headaches [30]. Furthermore, activation of the trigeminovascular system results in depletion of calcitonin gene-related peptide in the trigeminal ganglion [18] and trigeminal nucleus caudalis [21, 31, 32]. Thus, calcitonin gene-related peptide is released from the peripheral endings of sensory nerve fibers when the trigeminal system is activated, initiating a cascade of events which results in neurogenic inflammation. So far, there is no evidence that receptors for calcitonin gene-related peptide directly modulate dural neurogenic inflammation. In one report, the calcitonin gene-related peptide receptor antagonist, α-calcitonin gene-related peptide 8-37, inhibited dural neurogenic inflammation in guinea pigs but not in rats [33]. The development of novel compounds that bind with high affinity to the receptors for calcitonin gene-related peptide [34] is needed to further investigate their role in dural neurogenic inflammation.

Substance P

Receptor

Substance P is one of a group of related peptides known as tachykinins, which also include neurokinin A, neurokinin B, neuropeptide K and neuropeptide γ. The peptides share a common C-terminal sequence, which is required for most of their biological effects. They are encoded by two distinct genes, called pre-protachykinin A (substance P, neurokinin A, neuropeptide K and neuropeptide γ) and pre-protachykinin B (neurokinin B). The tachykinins exert their effects through one or more of three distinct G-protein coupled receptors, neurokinin-1 (substance P), neurokinin-2 (neurokinin A) and neurokinin-3 (neurokinin B) [35]. Auto-radiographic studies have shown the neurokinin-1 receptor, the target for substance P, to be widely distributed throughout the nervous system. They have been shown to be present on neurons of the trigeminal ganglion and trigeminal nucleus caudalis [36, 37], as well as on the peripheral endings of trigeminal sensory nerve fibers and endothelial cells [38, 39].

Co-localization

In the trigeminal ganglion and trigeminal nucleus caudalis, substance P is co-localized with several other neurotransmitters, in particular calcitonin gene-related peptide [18, 40, 41] and glutamate [42]. Like calcitonin gene-related peptide, substance P is released from the peripheral endings of trigeminal sensory nerve fibers [7, 43] around the dural blood vessels, inducing vasodilation [44]. The vasodilation induced by substance P is, however, short lived in comparison to that induced by calcitonin gene-related peptide. In addition, depletion of substance P has been found in the trigeminal nucleus caudalis, during or after activation of the trigeminovascular system [31, 32]. There is evidence that the synthesis of substance P is influenced by adrenal steroid hormones [45] and that release of the peptide is regulated through receptors for neurotrophic factor derived from glial cell line [46] and for N-methyl-D-aspartate (NMDA) [38, 47].

Nociception

The role of substance P in nociception is controversial [48]. Studies in knockout mice lacking the neurokinin-1 receptor suggest that the peptide is not involved in acute pain. However, it plays an important role in the control of spinal excitability, primarily through activation of central inhibitory pathways in the brainstem. Substance P was also found to be important for the survival of animals exposed to dangerous situations, inducing anxiety-like behavior. Animals lacking substance P and neurokinin A, in contrast, showed behavior that was remarkably different [49]. Additionally, clinical studies indicate a role for neurokinin-1-receptor antagonists not as analgesics but in the control of anxiety and, possibly, also depression [26, 50].

Inflammation

However, the involvement of substance P in inflammation has been well established. Substance P induces neurogenic inflammation in the dura mater, attenuated by neurokinin-1-receptor antagonists [51–54]. When added to human epidermis *in vitro*, it induced edema in a dose-dependent manner, with vasodilation and extravasation of lymphocytes and mast cells. Moreover, the release of inflammatory mediators, like interleukin 1α (IL-1α) and tumor necrosis factor (TNFα), was also observed, suggesting that substance P does indeed induce neurogenic inflammation in man [55].

Release

Hence, there is great synergy between calcitonin gene-related peptide and substance P in the production of dural neurogenic inflammation. However, the role of substance P as a pain modulator in the central nervous system is controversial. There is evidence that substance P is involved in neuronal activity of the trigeminal nucleus caudalis, as shown in studies using the c-fos marker [56–58], but not all studies agree [59]. Furthermore, when the trigeminal system is activated by electrical stimulation of the superior sagittal sinus, substance P is not increased in blood drawn from the internal jugular vein in cats [28]. It is important to mention that substance P is rapidly inactivated by specific peptidase enzymes, such as neutral endopeptidase-24, 11, acetylcholine esterase, cathepsin D and substance P-degrading enzyme [26, 60, 61], explaining why changes in the plasma level of substance P could sometimes not be detected. It has been shown that like calcitonin gene-related peptide, depletion of substance P in the trigeminal nucleus caudalis occurs after electrical stimulation of the trigeminal ganglion [62], presumably due to release of the peptide from the peripheral endings of the C-fibers. In a study in man, only calcitonin gene-related peptide and not substance P was found to be elevated in the plasma during migraine headache [29], whereas in another both peptides were found to be elevated [30]. After stimulation of the trigeminal ganglion, both substance P and calcitonin gene-related peptide are increased in blood drawn from the external jugular vein in man and animals [27]. Recent clinical studies using selective neurokinin-1-receptor antagonists failed to show efficacy in aborting migraine headache [63], raising doubts about the significance of substance P and, even more, of dural neurogenic inflammation in migraine pathogenesis [64]. However, the significance of substance P and the neurokinin-1 receptor as mediators of peripheral inflammation is well established. Additionally, recent imaging studies revealed preliminary evidence of dural neurogenic inflammation during migraine headache [65], making substance P deserving of further investigation in the pathogenesis of the migraine headache [66].

Serotonin

Receptors

Serotonin has been implicated in the pathogenesis of migraine since the early 1950s as an endogenous pain modulator and vasoactive substance [67]. In some but not in all studies [68], infusion of serotonin aborted spontaneous or reserpine-induced migraine headaches [69, 70]. Based in part on these observations, Humphrey et al developed the first serotonin-1B/1D-receptor agonist, sumatriptan, for the abortive treatment of migraine [71]. So far, 15 different serotonin receptors have been identified by pharmacological and molecular cloning techniques. They are grouped into seven classes according to transduction mechanisms and second messengers [24, 72]. Only a few of the serotonin-1 and -2 receptors, that is, the -1B, -1D, -1F and -2B receptors, are thought to be involved in pain transmission within the trigeminovascular system [73–76]. It is important to notice that the structure of the serotonin-1B receptor is different in man and rodents (a single amino acid difference in the seventh transmembrane domain), resulting in significant distinctions pharmacologically [77].

Location

The serotonin-1B, -D, and -F receptors have specific neurovascular distributions and functions within the trigeminovascular system. Immunohistochemical and *in situ* hybridization techniques co-localize the serotonin-1B receptor and its messenger RNA with substance P and calcitonin gene-related peptide in the trigeminal ganglion and trigeminal nucleus caudalis [78], suggesting that the receptor modulates the release of both peptides. *In vitro* studies by Durham et al have demonstrated that activation by sumatriptan of the serotonin-1B receptor on trigeminal ganglion cells is coupled to calcium-signaling pathways rather than cyclic AMP, leading to inhibition of the release of calcitonin gene-related peptide and transcription of the relevant gene [79–81]. Serotonin-1B-receptors and messenger RNA for the receptor (but not the serotonin-1D receptors) are densely localized in smooth muscle and less densely in endothelial cells of cerebral blood vessels [82, 83]. In human coronary arteries, the localization of serotonin-1B receptors is much less dense and has been shown on smooth muscle and is even less dense on endothelial cells [35, 84]. This particular vascular distribution of the receptor has been shown to determine the vasoconstrictor profile of the triptans [85–88]. The serotonin-1D receptor has been demonstrated in the trigeminal fibers and cell bodies, where its messenger RNA is co-localized only with messenger RNA encoding for calcitonin gene-related peptide. This suggests a possible role of the serotonin-1B receptor in inhibiting the release of calcitonin gene-related peptide but not of substance P [35, 89]. Like the serotonin-1B receptor, the 1F receptor is located on blood vessels, the trigeminal ganglion, and trigeminal nucleus caudalis [35, 90]. However, the serotonin-1F receptor lacks vasoconstrictor properties in isolated cerebral arteries, making it an attractive target for new antimigraine medications [91, 92]. A study indeed showed a selective serotonin-1F-receptor agonist to be effective in aborting migraine headache, although in high dosages only [93].

Activation

As is the case for sumatriptan, the triptans bind to the serotonin-1B, -D and -F receptors [94]. Sumatriptan inhibits, in a dose-dependent manner, plasma-protein extravasation induced by electrical stimulation of the trigeminal ganglion in guinea pigs [95] and so do the ergots and other triptans [10, 11, 13, 14, 96]. Studies in knockout mice lacking the serotonin-1B receptor revealed that sumatriptan inhibits plasma-protein extravasation *via* a mode of action not mediated by that receptor, suggesting that the serotonin-1D receptor may be more important in modulating dural neurogenic inflammation in the mouse [97]. Additionally, studies in several species have shown that sumatriptan inhibits plasma-protein extravasation *via* serotonin-1D receptors in guinea pigs and serotonin-1B receptors in rats [98]. Several selective serotonin-1D-receptor agonists were highly effective in inhibiting dural neurogenic inflammation in rats [98, 99] and guinea pigs [98–100]. Activation of the serotonin-1F receptor by potent selective agonists also showed high efficacy in inhibiting plasma-protein extravasation in rats and guinea pigs [101–103]. In other reports, the serotonin-1F-receptor agonist, LY-334370, had no effect on neurogenic dilation of dural blood vessels, whereas the compound was effective in inhibiting evoked potentials in the trigeminal nucleus caudalis. This indicates that the compound most likely has a central site of action on the second-order neurons [92]. However, the neurogenic vasodilation produced by electrical stimulation of the dura mater in rats is mediated *via* Aδ-fibers and not C-fibers [104]. It is not well characterized on which type of fiber the serotonin-1F receptor is located, but since LY-334370 was effective in inhibiting dural neurogenic inflammation in other studies [101], a peripheral site of action cannot be excluded. Thus, as with the triptans, the available data suggest a central as well as peripheral site of action for the serotonin-1F-receptor agonists. The selective serotonin-1D-receptor agonist, CP-122288, is highly effective in inhibiting dural neurogenic inflammation [97, 105, 106] but the compound was ineffective when tested clinically in migraineurs [107], again raising doubt about the significance of dural neurogenic inflammation in migraine. There is some evidence that the compound may also act on a different, as yet unknown receptor [108].

Other serotonin receptors

The role of other serotonin receptors in dural neurogenic inflammation has not been extensively investigated yet. The serotonin-1A receptor is involved in the inhibition of the synthesis, release and metabolism of serotonin [109]. Indirect evidence that the serotonin-1A receptor regulates dural neurogenic inflammation is based on the observation that dihydroergotamine, which non-selectively binds to this receptor with high affinity, inhibits dural neurogenic inflammation [110]. On the other hand, administration of the serotonin-1A-receptor agonist, 8-OH-DPAT, induced in serotonin-1B-receptor knockout mice a dose-dependent inhibition of serotonergic neuronal firing in the dorsal raphe nucleus, reversible with a serotonin-1A-receptor antagonist [111]. These findings suggest that the serotonin-1A receptor is involved in the modulation of sensory transmission in the trigeminal system, but further investigation is needed.

The role of the serotonin-2B receptor in migraine was also poorly characterized until recently. The serotonin-2B receptor has been shown to be involved in the modulation of cerebral blood flow during the vasodilation phase in cortical spreading depression, a phenomenon that may be related to the migraine aura [112]. The serotonin-2B receptor is located on the endothelial cells of cerebral blood vessels, coupled to nitric oxide synthase and promoting the local release of nitric oxide [112–114]. Nitric oxide is known to excite and sensitize the perivascular trigeminal nerve fibers through increasing cyclic GMP [115] and the release of neuropeptides [114, 116, 117]. Meta-chlorophenylpiperazine is a non-selective serotonin-2B-receptor agonist able to provoke migraine attacks in migraineurs, or headaches in normal individuals [118], and to induce dural neurogenic inflammation in animals [119]. RS-127445 is a selective, high affinity (pKi = 9.5), orally bioavailable serotonin-2B-receptor antagonist [73]. The compound has been shown to inhibit plasma-protein extravasation evoked by meta-chlorophenylpiperazine, as well as c-fos expression in the trigeminal nucleus caudalis evoked by capsaicin [120]. In other reports, selective serotonin-2B-receptor antagonists also inhibited dural neurogenic inflammation induced by meta-chlorophenylpiperazine [121]. Taken together, these data suggest that activation of serotonin-2B receptors is sufficient to generate neurogenic inflammation in the dura mater of the rat and to modulate c-fos response in second-order neurons following stimulation of primary sensory nerve fibers. Thus, like the serotonin-1B, -D and -F receptors, the serotonin-2B receptor seems to be important in migraine and related headaches and may serve as a target for future antimigraine medications.

Summary

In vivo studies demonstrated that dihydroergotamine, sumatriptan and avitriptan decrease plasma levels of calcitonin gene-related peptide in the superior sagittal sinus, induced by electrical stimulation of the trigeminal ganglion. This provides further evidence for modulation of the release of calcitonin gene-related peptide by serotonin-1B and -D receptors during dural neurogenic inflammation [11, 122]. Additionally, sumatriptan reduced the elevated plasma level of calcitonin gene-related peptide during migraine headache [123]. These findings, in turn, suggest that the role of serotonin in dural neurogenic inflammation may be indirect. It seems more likely that activation of serotonin-1 receptors decreases dural neurogenic inflammation mediated by calcitonin gene-related peptide by inhibiting the gene transcription and release of the peptide. Additionally, there is evidence that serotonin-1 receptors modulate pain transmission in the central nervous system as well, in particular in the trigeminal nucleus caudalis [124–127].

Nerve growth factor

Neurotrophic factor belongs to a family of soluble proteins mainly synthesized by nociceptive neurons. It plays a critical role in neuronal survival during development, while in adult neurons it regulates the growth and branching of sensory nerve fibers and neuropeptide expression in those involved in nociception. Nerve growth factor exerts its effects by high-affinity binding to the tyrosine-kinase receptor. Of particular interest has been the demonstration that nerve growth factor promotes the survival of small neurons in the trigeminal and dorsal root ganglia. Nerve growth factor also powerfully modulates the sensitivity of nociceptive nerve fibers to noxious stimuli and influences the generation of pain related to inflammation [128, 129]. Injections of nerve growth factor into the scalp caused the expression of messenger RNA for calcitonin gene-related peptide in trigeminal ganglion cells in guinea pigs [130]. Similarly, intraplanar injection of nerve growth factor into the rat hind paw increased the expression of messenger RNA for pre-pro-calcitonin gene-related peptide in the dorsal root ganglia [131]. The tyrosine-kinase A receptor for nerve growth factor has also been found by immunohistochemical techniques in nerve fibers innervating cranial blood vessels in man, which was co-localized with calcitonin gene-related peptide [132]. These data suggest that nerve growth factor regulates dural neurogenic inflammation by modulating the synthesis and release of calcitonin gene-related peptide and substance P [7, 133] in the trigeminal ganglion. The exact molecular pathway of this regulation is under investigation [134–136].

γ-aminobutyric acid

γ-Aminobutyric acid (GABA), the main inhibitory neurotransmitter in the central nervous system, exerts its effects by binding to two distinct receptors, A and B. The B receptor belongs to the protein-G coupled metabotropic receptors, whereas the A receptor is a chloride channel [137]. The first evidence for the involvement of GABA-related mechanisms in migraine pathogenesis came from the efficacy of the medication, valproate, in aborting migraine headache [138]. The medication was subsequently tested in animals and found to be effective in reducing the plasma-protein extravasation of dural neurogenic inflammation in guinea pigs, mediated by the GABA A receptor [15]. The neurosteroid, allopregnanolone, was studied in dural neurogenic inflammation in guinea pigs as well and showed high efficacy in reducing plasma-protein extravasation, also mediated by the A receptor [139], further indicating the significance of this receptor in modulating dural neurogenic inflammation. It is interesting, however, that this receptor has been detected only in the trigeminal nucleus caudalis [140] and not in peripheral parts of the trigeminal system. Peripheral A receptors are likely to be located on neurons of the parasympathetic sphenopalatine ganglion, since the inhibitory effect of A receptor agonists or modulators disappears in rats 2 weeks after ablation of the ganglion [15]. As with other receptors previously presented, the significance of the GABA A receptor is not limited to inhibiting dural neurogenic inflammation; the receptor is also involved in central pathways of orofacial pain transmission [141].

Other mediators of dural neurogenic inflammation

Prostaglandins

Early reports indicated that non-steroidal anti-inflammatory analgesics, including aspirin, were effective in inhibiting dural neurogenic inflammation, probably by modulating the release of substance P [9]. It is known that non-steroidal anti-inflammatory analgesics exert their anti-inflammatory effects by inhibition of peripheral prostaglandin synthesis, mainly by inhibiting the enzyme, cyclo-oxygenase 2 [142]. Prostaglandins, mostly prostaglandin E2 and prostacyclin, are important mediators of inflammation, pain and fever. They are synthesized in tissues by the constitutive enzyme, cyclo-oxygenase 1, and its isoform, cyclo-oxygenase 2, which can be induced with cytokines, growth factors, or other inflammatory stimuli. There is recent evidence that slowly developing pain, transmitted *via* C-fibers, is mediated by both cyclo-oxygenase 1 and 2, whereas rapidly transmitted pain is mediated only by cyclo-oxygenase 1 [143]. The significance of cyclo-oxygenase-2 at supraspinal levels in dural neurogenic inflammation in particular has not been investigated yet. The enzyme is mainly expressed in the trigeminal nucleus caudalis and in more central parts of the trigeminal system, but not in the cell bodies of the trigeminal ganglion [144–146]. There is recent evidence, however, that cyclo-oxygenase 2 is expressed during inflammation by endothelial brain cells, suggesting a possible role in dural neurogenic inflammation [147]. In one study, electrical stimulation of the trigeminal ganglion resulted in the release of calcitonin gene-related peptide and prostaglandins E2 from the dura mater [148]. Prostaglandin E2 binds with high affinity to the prostanoid receptor EP3, which has been detected on the cell bodies of the trigeminal ganglion [149]. These observations indicate an important role of the cyclooxygenase-prostaglandin system in dural neurogenic inflammation, which requires further investigation.

Other mediators

There is evidence that prejunctional α-adrenergic-2 and histamine-3 receptors, as well as somatostatin, inhibit dural neurogenic inflammation in guinea pigs and rats [52]. Flunarizine and novel selective calcium-channel blockers were also effective in modulating plasma-protein extravasation after electrical stimulation of the trigeminal ganglion [150]. Nitroglycerine is a nitric oxide donor and induces headache with migraine characteristics in migraineurs [151]. Nitric oxide is a vasoactive molecule that has been recently implicated in the pathogenesis of migraine [152, 153]. Apart from its vasodilating effects, there is evidence that nitric oxide is also involved in inflammation, by modulating the activity of cyclo-oxygenase [154]. Like meta-chlorophenylpiperazine and substance P, nitroglycerine induces dural neurogenic inflammation in animals, mediated by activation of nitric oxide synthase [119]. Nitric oxide sensitizes the perivascular trigeminal nerve fibers by releasing calcitonin gene-related peptide and substance P [116]. Recent experimental data suggest that nitroglycerine results in delayed inflammation in the rat dura mater, possibly corresponding to the delayed headache in man, *via* up-regulation of pro-inflammatory genes, such as NF-κB, a transcription factor that regulates the expression of the gene encoding for nitric oxide synthase 155].

Capsaicin

Capsaicin is an irritant with selective excitatory and toxic effects on damage-sensing peripheral nerve fibers. It binds to a receptor that has been recently cloned, the vanilloid 1 receptor [156, 157]. This receptor is a heat-sensitive ion channel located on the peripheral endings of sensory nerve fibers, mediating the response of small-diameter C-fibers. Capsaicin causes depolarization and firing of C-fibers mediated by mechanisms involving calcium and sodium, releasing calcitonin gene-related peptide and substance P, although the vanilloid 1 receptor is not co-localized in nerve fibers containing substance P [19, 158]. This receptor is also activated by noxious heat as well as by the low pH often associated with tissue damage and inflammation. It is interesting that low pH in combination with bradykinin has been found to be the most effective chemical stimulus for activating dural nociceptors [159].

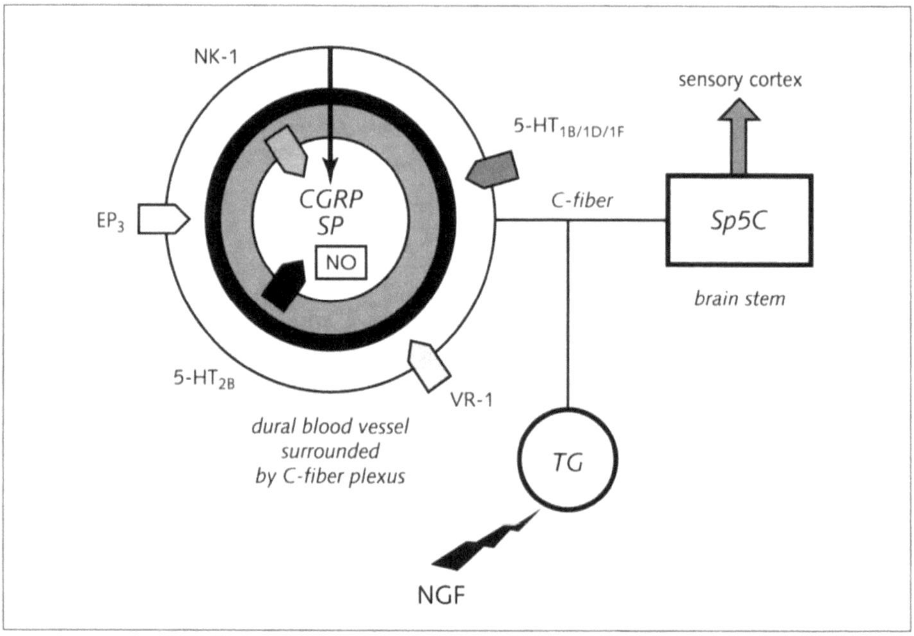

Figure 1
Head pain is transmitted via the trigeminal nerve, which contains several types of fibers of which the unmyelinated C-fibers are nociceptive. They originate from the trigeminal ganglion (TG) and are marked by the expression of calcitonin gene-related peptide (CGRP), substance P (SP), their receptors, and receptors for nerve growth factor (NGF). Their peripheral endings surround the cranial blood vessels while their central endings terminate in the trigeminal nucleus caudalis (Sp5C). From the trigeminal nucleus caudalis, the pain is conducted to other brainstem nuclei and higher cortical structures for registration and modulation of nociceptive information. Activation of serotonin-1D ($5\text{-}HT_{1D}$), -1B ($5\text{-}HT_{1B}$) and -1F ($5\text{-}HT_{1F}$) receptors and blockade of serotonin-2B ($5\text{-}HT_{2B}$) and neurokinin-1 (NK-1) receptors inhibit dural neurogenic inflammation. When serotonin-2B receptors are activated, nitric oxide (NO) is released from endothelial cells and induces dural neurogenic inflammation.

Conclusion

Dural neurogenic inflammation occurs after stimulation of the trigeminovascular system and compounds that inhibit dural neurogenic inflammation in animals may be effective in aborting migraine headache. With the exception of opioids, which mostly act centrally, all abortive antimigraine medications are effective in modulating dural neurogenic inflammation. However, not all compounds that inhibit dural neurogenic inflammation have clinical importance (for example, serotonin-1D-receptor agonists and neurokinin-1-receptor antagonists), raising arguments against the significance of dural neurogenic inflammation in migraine. Recent neuroimaging revealed preliminary evidence for dural neurogenic inflammation during migraine headache, however. In animals, dural neurogenic inflammation is induced when the trigeminovascular system is activated electrically, mechanically, or chemically, the latter with compounds that recognize specific receptors on the vascular endothelial cells. Calcitonin gene-related peptide and substance P are the most likely peptides that directly modulate dural neurogenic inflammation (Fig. 1). There is evidence that other modulators are also involved in dural neurogenic inflammation, like nitric oxide, but further studies are needed to clarify their specific molecular pathways. Release of calcitonin gene-related peptide and substance P is modulated by several receptors as shown in Figure 1. Specific agonists or antagonists of these receptors inhibit dural neurogenic inflammation, predicting that they may have antimigraine efficacy in man.

References

1 Moskowitz MA, Reinhard JF, Romero J et al (1979) Neurotransmitters and the fifth cranial nerve: is there a relation to the headache phase of migraine? *Lancet* 2: 883–885

2 McMahon SB, Bennett GJ (2000) Glial cell-derived neurotrophic factor and nociceptive neurons. In: JN Wood (ed): *Molecular basis of pain induction.* Wiley-Liss, New York, 65–86

3 Moskowitz MA (1993) Neurogenic inflammation in the pathophysiology and treatment of migraine. *Neurology* 43: S16–S20

4 Markowitz S, Saito K, Moskowitz MA (1988) Neurogenically mediated plasma extravasation in dura mater: effect of ergot alkaloids. A possible mechanism of action in vascular headache. *Cephalalgia* 8: 83–91

5 Markowitz S, Saito K, Moskowitz MA (1987) Neurogenically mediated leakage of plasma protein occurs from blood vessels in dura mater but not brain. *J Neurosci* 7: 4129–4136

6 Dimitriadou V, Buzzi MG, Theoharides TC, Moskowitz, MA (1992) Ultrastructural evidence for neurogenically mediated changes in blood vessels of the rat dura mater and tongue following antidromic trigeminal stimulation. *Neuroscience* 48: 187–203

7 Rozniecki JJ, Dimitriadou V, Lambracht-Hall M et al (1999) Morphological and functional demonstration of rat dura mater mast cell-neuron interactions *in vitro* and *in vivo*. *Brain Res* 49: 1–15

8 Delepine L, Aubineau P (1997) Plasma protein extravasation induced in the rat dura mater by stimulation of the parasympathetic sphenopalatine ganglion. *Exp Neurol* 147: 389–400

9 Buzzi MG, Sakas DE, Moskowitz MA (1989) Indomethacin and acetylsalicylic acid

block neurogenic plasma protein extravasation in rat dura mater. *Eur J Pharmacol* 165: 251–258

10 Saito K, Markowitz S, Moskowitz MA (1988) Ergot alkaloids block neurogenic extravasation in dura mater: proposed action in vascular headaches. *Ann Neurol* 24: 732–737

11 Buzzi MG, Carter WB, Shimizu T et al (1991) Dihydroergotamine and sumatriptan attenuate levels of CGRP in plasma in rat superior sagittal sinus during electrical stimulation of the trigeminal ganglion. *Neuropharmacology* 30: 1193–1200

12 Martin GR, Robertson AD, MacLennan SJ et al (1997) Receptor specificity and trigemino-vascular inhibitory actions of a novel $5\text{-HT}_{1B/1D}$ receptor partial agonist, 311C90 (zolmitriptan). Br J Pharmacol 121: 157–164

13 Connor HE, Feniuk W, Beattie DT et al (1997) Naratriptan: biological profile in animal models relevant to migraine. *Cephalalgia* 17: 145–152

14 Williamson DJ, Shepheard SL, Hill RG, Hargreaves RJ (1997) The novel anti-migraine agent rizatriptan inhibits neurogenic dural vasodilation and extravasation. *Eur J Pharmacol* 328: 61–64

15 Lee WS, Limmroth V, Ayata C et al (1995) Peripheral GABA-A receptor-mediated effects of sodium valproate on dural plasma protein extravasation to substance P and trigeminal stimulation. *Br J Pharmacol* 116: 1661–1667

16 Olivar T, Razzaque Z, Nwagwu M, Longmore J (2000) Neurogenic vasodilation in rabbit basilar isolated artery: involvement of calcitonin gene-related peptide. *Eur J Pharmacol* 395: 61–68

17 Edvinsson L, Ekman R, Jansen I et al (1987) Calcitonin gene-related peptide and cerebral blood vessels: distribution and vasomotor effects. *J Cereb Blood Flow Metabol* 7: 720–728

18 Hutchins B, Spears R, Hinton RJ, Harper RP (2000) Calcitonin gene-related peptide and substance P immunoreactivity in rat trigeminal ganglia and brainstem following adjuvant-induced inflammation of the temporomandibular joint. *Arch Oral Biol* 45: 335–345

19 Spears R, Hutchins B, Hinton RJ (1998) Capsaicin application to the temporomandibular joint alters calcitonin gene-related peptide levels in the trigeminal ganglion of the rat. *J Orofac Pain* 12: 108–115

20 Ichikawa H, Sugimoto T (1999) Peptide 19-immunoreactive primary sensory neurons in the rat trigeminal ganglion. *Brain Res* 846: 274–279

21 Knyihar-Csillik E, Tajti J, Samsam M et al (1998) Depletion of calcitonin gene-related peptide from the caudal trigeminal nucleus of the rat after electrical stimulation of the Gasserian ganglion. *Exp Brain Res* 118: 111–114

22 Van Rossum D, Hanisch UK, Quirion R (1997) Neuroanatomical localization, pharmacological characterization and functions of CGRP, related peptides and their receptors. *Neurosci Biobehav Rev* 21: 649–678

23 Moreno MJ, Cohen Z, Stanimirovic DB, Hamel E (1999) Functional calcitonin gene-related peptide type 1 and adrenomedullin receptors in human trigeminal ganglia, brain

vessels, and cerebromicrovascular or astroglial cells in culture. *J Cereb Blood Flow Metabol* 19: 1270–1278
24 Alexander SPH, Peters JA (1997) Receptor and ion channel nomenclature. *Trends Pharmacol Sci* (Suppl 1) 18: 1–82
25 Steinhoff M, Vergnolle N, Young SH et al (2000) Agonists of proteinase-activated receptor 2 induce inflammation by a neurogenic mechanism. *Nat Med* 6: 151–158
26 Hunt SP, O'Brien JA, Palmer JA (2000) Role of substance P in nociception, analgesia, and aggression. In: JN Wood (ed): *Molecular basis of pain induction*. Wiley-Liss, New York, 209–260
27 Goadsby PJ, Edvinsson L, Ekman R (1988) Release of vasoactive peptides in the extracerebral circulation of humans and the cat during activation of the trigeminovascular system. *Ann Neurol* 23: 193–196
28 Zagami AS, Goadsby PJ, Edvinsson L (1990) Stimulation of the superior sagittal sinus in the cat causes release of vasoactive peptides. *Neuropeptides* 16: 69–75
29 Goadsby PJ, Edvinsson L, Ekman R (1990) Vasoactive peptide release in the extracerebral circulation of humans during migraine headache. *Ann Neurol* 28: 183–187
30 Gallai V, Sarchielli P, Floridi A et al (1995) Vasoactive peptide levels in the plasma of young migraine patients with and without aura assessed both interictally and ictally *Cephalalgia* 15: 384–390
31 Samsam M, Covenas R, Ahangari R et al (1999) Simultaneous depletion of neurokinin A, substance P and calcitonin gene-related peptide from the caudal trigeminal nucleus of the rat during electrical stimulation of the trigeminal ganglion. *Pain* 84: 389–395
32 Samsam M, Covenas R, Ahangari Ret al (1999) Alterations in neurokinin A, substance P and calcitonin gene-related peptide immunoreactivities in the caudal trigeminal nucleus of the rat following electrical stimulation of the trigeminal ganglion. *Neurosci Lett* 261: 179–182
33 O'Shaughnessy CT, Connor HE (1994) Investigation of the role of tachykinin NK1, NK2 receptors and CGRP receptors in neurogenic plasma protein extravasation in dura mater. *Eur J Pharmacol* 263: 193–198
34 Doods H, Hallermayer G, Wu D et al (2000) Pharmacological profile of BIBN4096BS, the first selective small molecule CGRP antagonist. *Br J Pharmacol* 129: 420–423
35 Longmore J, Shaw D, Smith D et al (1997) Differential distribution of $5HT_{1D}$- and $5HT_{1B}$-immunoreactivity within the human trigemino-cerebrovascular system: implications for the discovery of new antimigraine drugs. *Cephalalgia* 17: 833–842
36 Patacchini R, Maggi CA (1995) Tachykinin receptors and receptor subtypes. *Arch Int Pharmacodyn Ther* 329: 161–184
37 Matthews MA, Hoffmann KD, Stover JD (1992) Ultrastructural characterization of substance-P-immunoreactive synaptic terminals in the cat's normal and rhizotomized trigeminal subnucleus caudalis. *Somatosens Mot Res* 9: 131–156
38 Carlton SM, Zhou S, Coggeshall RE (1998) Evidence for the interaction of glutamate and NK1 receptors in the periphery. *Brain Res* 790: 160–169
39 Liu-Chen LY, Liszczak TM, King JC, Moskowitz MA (1986) Immunoelectron micro-

scopic study of substance P-containing fibers in feline cerebral arteries. *Brain Res* 369: 12–20
40 Matsuda H, Kusakabe T, Hayashida Y et al (1998) Substance P- and calcitonin gene-related peptide-containing nerve fibers in the nasal mucosa of chronically hypoxic rats. *Brain Res Bull* 45: 563–569
41 Gibbins IL, Furness JB, Costa M et al (1985) Co-localization of calcitonin gene-related peptide-like immunoreactivity with substance P in cutaneous, vascular and visceral sensory neurons of guinea pigs. *Brain Res Bull* 57: 125–130
42 Battaglia G, Rustioni A (1988) Coexistence of glutamate and substance P in dorsal root ganglion neurons of the rat and monkey. *J Comp Neurol* 277: 302–312
43 Moskowitz MA, Brody M, Liu-Chen LY (1983) *In vitro* release of immunoreactive substance P from putative afferent nerve endings in bovine pia arachnoid. *Neuroscience* 9: 809–814
44 Reid J, McCulloch J (1987) Capsaicin and blood-brain barrier permeability. *Neurosci Lett* 81: 165–170
45 Covenas R, DeLeon M, Chadi G et al (1994) Adrenalectomy increases the number of substance P and somatostatin immunoreactive nerve cells in the rat lumbar dorsal root ganglia. *Brain Res* 640: 352–356
46 Ogun-Muyiwa P, Helliwell R, McIntyre P, Winter J (1999) Glial cell line-derived neurotrophic factor (GDNF) regulates VR1 and substance P in cultured sensory neurons. *Neuroreport* 10: 2107–2111
47 Liu H, Mantyh PW, Basbaum AI (1997) NMDA-receptor regulation of substance P release from primary afferent nociceptors. *Nature* 386: 721–724
48 Iversen L (1998) Substance P equals pain substance? *Nature* 392: 334–335
49 De Felipe C, Herrero JF, O'Brien JA et al (1998) Altered nociception, analgesia and aggression in mice lacking the receptor for substance P. *Nature* 392: 394–397
50 Kramer MS, Cutler N, Feighner J et al (1998) Distinct mechanism for antidepressant activity by blockade of central substance P receptors. *Science* 281: 1640–1645
51 Beattie DT, Beresford IJ, Connor HE et al (1995) The pharmacology of GR203040, a novel, potent and selective non-peptide tachykinin NK1 receptor antagonist. *Br J Pharmacol* 116: 3149–3157
52 Matsubara T, Moskowitz MA, Huang Z (1992) UK-14,304, R(–)-alpha-methyl-histamine and SMS 201-995 block plasma protein leakage within dura mater by prejunctional mechanisms. *Eur J Pharmacol* 224: 145–150
53 Polley JS, Gaskin PJ, Perren MJ et al (1997) The activity of GR205171, a potent non-peptide tachykinin NK1 receptor antagonist, in the trigeminovascular system. *Regul Pept* 68: 23–29
54 Michaud JC, Alonso R, Gueudet C et al (1998) Effects of SR140333, a selective non-peptide NK1 receptor antagonist, on trigemino-thalamic nociceptive pathways in the rat. *Fundam Clin Pharmacol* 12: 88–94
55 Branchet-Gumila MC, Boisnic S, Le Charpentier Y et al (1999) Neurogenic modifica-

tions induced by substance P in an organ culture of human skin. *Skin Pharmacol Appl Skin Physiol* 12: 211–220
56 Bereiter DA, Bereiter DF, Tonnessen BH, MacLean DB (1998) Selective blockade of substance P or neurokinin A receptors reduces the expression of c-fos in trigeminal subnucleus caudalis after corneal stimulation in the rat. *Neuroscience* 83: 525–534
57 Clayton JS, Gaskin PJ, Beattie DT (1997) Attenuation of fos-like immunoreactivity in the trigeminal nucleus caudalis following trigeminovascular activation in the anaesthetised guinea-pig. *Brain Res* 775: 74–80
58 Cutrer FM, Moussaoui S, Garret C, Moskowitz MA (1995) The non-peptide neurokinin-1 antagonist, RPR 100893, decreases c-fos expression in trigeminal nucleus caudalis following noxious chemical meningeal stimulation. *Neuroscience* 64: 741–750
59 Goadsby PJ, Hoskin KL, Knight YE (1998) Substance P blockade with the potent and centrally acting antagonist GR205171 does not effect central trigeminal activity with superior sagittal sinus stimulation. *Neuroscience* 86: 337–343
60 Iversen LL, Jessell T, Kanazawa I (1976) Release and metabolism of substance P in rat hypothalamus. *Nature* 264: 81–83
61 Hall ME, Miley F, Stewart JM (1989) The role of enzymatic processing in the biological actions of substance P. *Peptides* 10: 895–901
62 Samsam M, Covenas R, Ahangari R et al (2000) Simultaneous depletion of neurokinin A, substance P and calcitonin gene-related peptide from the caudal trigeminal nucleus of the rat during electrical stimulation of the trigeminal ganglion. *Pain* 84: 389–395
63 Goldstein DJ, Wang O, Saper JR et al (1997) Ineffectiveness of neurokinin-1 antagonist in acute migraine: a crossover study. *Cephalalgia* 17: 785–790
64 May A, Shepheard SL, Knorr M et al (1998) Retinal plasma extravasation in animals but not in humans: implications for the pathophysiology of migraine. *Brain* 121: 1231–1237
65 Pappagalo M, SzaboZ, Esposito G et al (1999) Imaging neurogenic inflammation in patients with migraine headaches (abstract). *Neurology* 52: A274–A275
66 Moskowitz MA, Mitsikostas DD (1997) A negative clinical study in the search for a migraine treatment (editorial). *Pain* 17: 720–721
67 Wolff HG (1963) *Headache and other headache pain*. Oxford University Press, New York
68 Ostfeld AM (1959) Some aspects of cardiovascular regulation in man. *Angiology* 10: 34–42
69 Kimball RW, Friedman AP, Vallejo E (1960) Effect of serotonin in migraine patients. *Neurology* 10: 107–111
70 Anthony M, Hinterberger H, Lance LW (1969) The possible relationship of serotonin to the migraine syndrome. *Res Clin Stud Headache* 2: 29–59
71 Humphrey PPA, Feniuk W, Marriott AS et al (1991) Preclinical studies on the anti-migraine drug, sumatriptan. *Eur Neurol* 31: 282–290
72 Hoyer D, Martin G (1997) 5-HT receptor classification and nomenclature: towards a harmonization with the human genome. *Neuropharmacology* 36: 419–428

73 Bonhaus DW, Flippin LA, Greenhouse RJ et al (1999) RS-127445: a selective, high affinity, orally bioavailable 5-HT_{2B} receptor antagonist. *Br J Pharmacol* 127: 1075–1082

74 Goadsby PJ, Hoskin KL (1998) Serotonin inhibits trigeminal nucleus activity evoked by craniovascular stimulation through a $5HT_{1B/1D}$ receptor: a central action in migraine? *Ann Neurol* 43: 711–718

75 Johnson KW, Phebus LA, Cohen ML (1998) Serotonin in migraine: theories, animal models and emerging therapies. *Prog Drug Res* 51: 219–244

76 Mitsikostas DD, Moskowitz MA (1998) Serotonin's central action in migraine. *Neurology Network Commentary* 2: 232–235

77 Oksenberg D, Marsters SA, O'Dowd BF et al (1992) A single amino-acid difference confers major pharmacological variation between human and rodent 5-HT_{1B} receptors. *Nature* 360: 161–163

78 Bruinvels AT, Landwehrmeyer B, Moskowitz MA, Hoyer D (1992) Evidence for the presence of 5-HT_{1B} receptor messenger RNA in neurons of the rat trigeminal ganglia. *Eur J Pharmacol* 227: 357–359

79 Durham PL, Russo AF (1998) Serotonergic repression of mitogen-activated protein kinase control of the calcitonin gene-related peptide enhancer. *Mol Endocrinol* 12: 1002–1009

80 Durham PL, Sharma RV, Russo AF (1997) Repression of the calcitonin gene-related peptide promoter by 5-HT_1 receptor activation. *J Neurosci* 17: 9545–9553

81 Durham PL, Russo AF (1999) Regulation of calcitonin gene-related peptide secretion by a serotonergic antimigraine drug. *J Neurosci* 19: 3423–3429

82 Bouchelet I, Cohen Z, Case B et al (1996) Differential expression of sumatriptan-sensitive 5-hydroxytryptamine receptors in human trigeminal ganglia and cerebral blood vessels. *Mol Pharmacol* 50: 219–223

83 Nilsson T, Longmore J, Shaw D et al (1999) Contractile 5-HT_{1B} receptors in human cerebral arteries: pharmacological characterization and localization with immunocytochemistry. *Br J Pharmacol* 128: 1133–1140

84 Nilsson T, Longmore J, Shaw D et al (1999) Characterisation of 5-HT receptors in human coronary arteries by molecular and pharmacological techniques. *Eur J Pharmacol* 372: 49–56

85 Hamel E, Gregoire L, Lau B (1993) 5-HT_1 receptors mediating contraction in bovine cerebral arteries: a model for human cerebrovascular '5-HT_{1D} beta' receptors. *Eur J Pharmacol* 242: 75–82

86 Kaumann AJ, Parsons AA, Brown AM (1993) Human arterial constrictor serotonin receptors. *Cardiovasc Res* 27: 2094–2103

87 Lambert G, Michalicek J (1996) Effect of antimigraine drugs on dural blood flows and resistances and the responses to trigeminal stimulation. *Eur J Pharmacol* 311: 141–151

88 Longmore J, Maguire JJ, MacLeod A et al (2000) Comparison of the vasoconstrictor effects of the selective 5-HT_{1D}-receptor agonist L-775,606 with the mixed 5-$HT_{1B/1D}$-receptor agonist sumatriptan and 5-HT in human isolated coronary artery. *Br J Clin Neurosci Lett* 49: 126–131

89 Bonaventure P, Langlois X, Leysen JE (1998) Co-localization of 5-HT_{1B}- and 5-HT_{1D} receptor mRNA in serotonergic cell bodies in guinea pig dorsal raphe nucleus: a double labeling *in situ* hybridization histochemistry study. *Neurosci Lett* 254: 113–116

90 Adham N, Bard JA, Zgombick JM et al (1997) Cloning and characterization of the guinea pig 5-HT_{1F} receptor subtype: a comparison of the pharmacological profile to the human species homolog. *Neuropharmacology* 36: 569–576

91 Razzaque Z, Heald MA, Pickard JD et al (1999) Vasoconstriction in human isolated middle meningeal arteries: determining the contribution of 5-HT_{1B}- and 5-HT_{1F}-receptor activation. *Br J Clin Pharmacol* 47: 75–82

92 Shepheard S, Edvinsson L, Cumberbatch M et al (1999) Possible antimigraine mechanisms of action of the 5HT_{1F} receptor agonist LY334370. *Cephalalgia* 19: 851–858

93 Goldstein, DJ, Roon KI, Offen WW et al (1999) Migraine treatment with selective 5-HT_{1F} receptor agonist (SSOFRA) LY334370 (abstract). *Cephalalgia* 19: 318

94 Waeber C, Moskowitz MA (1995) [3H]sumatriptan labels both 5-HT_{1D} and 5-HT_{1F} receptor binding sites in the guinea pig brain: an autoradiographic study. *Naunyn Schmiedebergs Arch Pharmacol* 352: 263–275

95 Buzzi MG, Moskowitz MA (1990) The antimigraine drug, sumatriptan (GR43175), selectively blocks neurogenic plasma extravasation from blood vessels in dura mater. *Br J Pharmacol* 99: 202–206

96 Martin GR, Robertson AD, MacLennan SJ et al (1997) Receptor specificity and trigemino-vascular inhibitory actions of a novel 5-$HT_{1B/1D}$ receptor partial agonist, 311C90 (zolmitriptan). *Br J Pharmacol* 121: 157–164

97 Yu XJ, Waeber C, Castanon N et al (1996) 5-Carboxamido-tryptamine, CP-122,288 and dihydroergotamine but not sumatriptan, CP-93,129, and serotonin-5-O-carboxymethyl-glycyl-tyrosinamide block dural plasma protein extravasation in knockout mice that lack 5-hydroxytryptamine1B receptors. *Mol Pharmacol* 49: 761–765

98 Yu XJ, Cutrer FM, Moskowitz MA, Waeber C (1997) The 5-HT_{1D} receptor antagonist GR-127,935 prevents inhibitory effects of sumatriptan but not CP-122,288 and 5-CT on neurogenic plasma extravasation within guinea pig dura mater. *Neuropharmacology* 36: 83–91

99 Petty MA, Elands J, Johnson MP et al (1997) The selectivity of MDL 74,721 in models of neurogenic *versus* vascular components of migraine. *Eur J Pharmacol* 336: 127–136

100 Cutrer FM, Yu XJ, Ayata G et al (1999) Effects of PNU-109,291, a selective 5-HT_{1D} receptor agonist, on electrically induced dural plasma extravasation and capsaicin-evoked c-fos immunoreactivity within trigeminal nucleus caudalis. *Neuropharmacology* 38: 1043–1053

101 Wainscott DB, Johnson KW, Phebus LA et al (1998) Human 5-HT_{1F} receptor-stimulated [35S]GTPgammaS binding: correlation with inhibition of guinea pig dural plasma protein extravasation. *Eur J Pharmacol* 352: 117–124

102 Phebus LA, Johnson KW, Zgombick JM et al (1997) Characterization of LY344864 as a pharmacological tool to study 5-HT_{1F} receptors: binding affinities, brain penetration

and activity in the neurogenic dural inflammation model of migraine. *Life Sci* 61: 2117–2126
103 Johnson KW, Schaus JM, Durkin MM et al (1997) 5-HT_{1F} receptor agonists inhibit neurogenic dural inflammation in guinea pigs. *Neuroreport* 8: 2237–2240
104 Shepherd SL, Williamson DJ, Beer MS et al (1997) Differential effects of 5-$HT_{1B/1D}$ receptor agonists on neurogenic dural plasma extravasation and vasodilation in anaesthetized rats. *Neuropharmacolgy* 36: 525–533
105 Gupta P, Brown D, Butler P et al (1995) The *in vivo* pharmacological profile of a 5-HT_1 receptor agonist, CP-122,288, a selective inhibitor of neurogenic inflammation. *Br J Pharmacol* 116: 2385–2390
106 Lee WS, Moskowitz MA (1993) Conformationally restricted sumatriptan analogues, CP-122,288 and CP- 122,638, exhibit enhanced potency against neurogenic inflammation in dura mater. *Brain Res* 626: 303–305
107 Roon KI, Olesen J, Diener HC et al (2000) No acute antimigraine efficacy of CP-122,288, a highly potent inhibitor of neurogenic inflammation: results of two randomized, double-blind, placebo-controlled clinical trials. *Ann Neurol* 47: 238–241
108 Martin G (1997) Serotonin receptor involvement in the pathogenesis and treatment of migraine. In: PJ Goadsby, SD Silberstein (eds): *Headache*. Butterworth-Heinemann, New York, 25–38
109 Peroutka SJ (1988) Antimigraine drug interactions with serotonin receptor subtypes in human brain. *Ann Neurol* 23: 500–504
110 Buzzi MG, Moskowitz MA, Peroutka SJ, Byun B (1991) Further characterization of the putative 5-HT receptor which mediates blockade of neurogenic plasma extravasation in rat dura mater. *Br J Pharmacol* 103: 1421–1428
111 Evrard A, Laporte AM, Chastanet M et al (1999) 5-HT_{1A} and 5-HT_{1B} receptors control the firing of serotoninergic neurons in the dorsal raphe nucleus of the mouse: studies in 5-HT_{1B} knock-out mice. *Eur J Neurosci* 11: 3823–3831
112 Schmuck K, Ullmer C, Kalkman HO et al (1996) Activation of meningeal 5-HT_{2B} receptors: an early step in the generation of migraine headache? *Eur J Neurosci* 8: 959–967
113 Martin GR, Bolofo ML, Giles H (1992) Inhibition of endothelium-dependent vasorelaxation by arginine analogues: a pharmacological analysis of agonist and tissue dependence. *Br J Pharmacol* 105: 643–652
114 Wei EP, Moskowitz MA, Boccalini P, Kontos HA (1992) Calcitonin gene-related peptide mediates nitroglycerin and sodium nitroprusside-induced vasodilation in feline cerebral arterioles. *Circ Res* 70: 1313–1319
115 Aley KO, McCarter G, Levine JD (1998) Nitric oxide signaling in pain and nociceptor sensitization in the rat. *J Neurosci* 18: 7008–7014
116 Garry MG, Richardson JD, Hargreaves KM (1994) Sodium nitroprusside evokes the release of immunoreactive calcitonin gene-related peptide and substance P from dorsal horn slices *via* nitric oxide-dependent and nitric oxide-independent mechanisms. *J Neurosci* 14: 4329–4337

117 Yonehara, N, Yoshimura M (1999) Effect of nitric oxide on substance P release from the peripheral endings of primary afferent neurons. *Neurosci Lett* 271: 199–201
118 Silberstein SD, Fozard JR, Murphy DL (1992) More on mCPP and migraine (letter). *Headache* 32: 242–244
119 Johnson KW, Nelson DB, Wainscott DB et al (1997) mCPP-induced dural extravasation-a potential model of migraine prophylaxis [abstract]. *Cephalalgia* 17: 342
120 Bonhaus DW, Chang LK, Cao Z et al (1999) RS-127445, a selective 5-HT_{2B} receptor antagonist, blocks mCPP-induced plasma protein extravasation in dura mater and capsaicin-evoked c-fos expression in trigeminal nucleus caudalis. In: J Olesen, PJ Goadsby (eds): *Cluster headache and related headaches.* Oxford University Press, New York, 278–286
121 Nelson AD, Wainscott DB, Lucaites VL et al (1997) Selective $5HT_{2B}$ receptor antagonists block mCPP-induced dural extravasation (abstract). *Cephalalgia* 17: 342
122 Knight YE, Edvinsson L, Goadsby PJ (1999) Blockade of calcitonin gene-related peptide release after superior sagittal sinus stimulation in cat: a comparison of avitriptan and CP122,288. *Neuropeptides* 33: 41–46
123 Goadsby PJ, Edvinsson L (1993) The trigeminovascular system and migraine: studies characterizing cerebrovascular and neuropeptide changes seen in humans and cats. *Ann Neurol* 33: 48–56
124 Kaube H, Hoskin KL, Goadsby PJ (1993) Inhibition by sumatriptan of central trigeminal neurons only after blood-brain barrier disruption. *Br J Pharmacol* 109: 788–792
125 Shepheard SL, Williamson DJ, Williams J et al (1995) Comparison of the effects of sumatriptan and the NK1 antagonist CP- 99,994 on plasma extravasation in dura mater and c-fos mRNA expression in trigeminal nucleus caudalis of rats. *Neuropharmacology* 34: 255–261
126 Goadsby PJ, Hoskin KL (1996) Inhibition of trigeminal neurons by intravenous administration of the serotonin $(5HT)_{1B/D}$ receptor agonist zolmitriptan (311C90): are brain stem sites therapeutic target in migraine? *Pain* 67: 355–359
127 Storer RJ, Goadsby PJ (1997) Microiontophoretic application of serotonin $(5HT)_{1B/1D}$ agonists inhibits trigeminal cell firing in the cat. *Brain* 120: 2171–2177
128 Lewin GR, Mendell LM (1993) Nerve growth factor and nociception. *Trends Neurosci* 16: 353–359
129 Woolf CJ, Safieh-Garabedian B, Ma QP et al (1994) Nerve growth factor contributes to the generation of inflammatory sensory hypersensitivity. *Neuroscience* 62: 327–331
130 Schicho R, Skofitsch G, Donnerer J (1999) Regenerative effect of human recombinant NGF on capsaicin-lesioned sensory neurons in the adult rat. *Brain Res* 815: 60–69
131 Amann R, Sirinathsinghji DJ, Donnerer J et al (1996) Stimulation by nerve growth factor of neuropeptide synthesis in the adult rat *in vivo*: bilateral response to unilateral intraplantar injections. *Neurosci Lett* 203: 171–174
132 Saldanha G, Hongo J, Plant G et al. (1999) Decreased CGRP but preserved Trk A immunoreactivity in nerve fibers in inflamed human superficial temporal arteries. *J Neurol Neurosurg Psychiatry* 66: 390–392

133 Fernyhough P, Brewster WJ, Fernandes K et al (1998) Stimulation of nerve growth-factor and substance P expression in the iris-trigeminal axis of diabetic rats – involvement of oxidative stress and effects of aldose reductase inhibition. *Brain Res* 802: 247–253
134 Freeland K, Liu YZ, Latchman DS (2000) Distinct signaling pathways mediate the cAMP response element (CRE)-dependent activation of the calcitonin gene-related peptide gene promoter by cAMP and nerve growth factor. *Biochem J* 345: 233–238
135 Supowit SC, Christensen MD, Westlund KN et al (1995) Dexamethason and activators of the protein kinase A and C signal transduction pathways regulate neuronal calcitonin gene-related peptide expression and release. *Brain Res* 686: 77–86
136 Watson A, Latchman D (1995) The cyclic AMP response element in the calcitonin/calcitonin gene-related peptide gene promoter is necessary but not sufficient for its activation by nerve growth factor. *J Biol Chem* 270: 9655–9660
137 Bormann I (2000) The 'ABC' of GABA receptors. *Trends Pharmacol Sci* 21: 16–19
138 Jensen R, Brinck T, Olesen J (1994) Sodium valproate has a prophylactic effect in migraine without aura: a triple-blind, placebo-controlled crossover study. *Neurology* 44: 647–651
139 Cutrer FM, Moskowitz MA (1996) The actions of valproate and neurosteroids in a model of trigeminal pain. *Headache* 36: 579–585
140 Dumba JS, Irish PS, Anderson NL, Westrum LE (1998) Electron microscopic analysis of gamma-aminobutyric acid and glycine colocalization in rat trigeminal subnucleus caudalis. *Brain Res* 806: 16–25
141 Cutrer FM, Limmroth V, Ayata G, Moskowitz MA (1995) Attenuation by valproate of c-fos immunoreactivity in trigeminal nucleus caudalis induced by intracisternal capsaicin. *Br J Pharmacol* 116: 3199–3204
142 Cashman JN (1996) The mechanisms of action of NSAIDs in analgesia. *Drugs* 52 (Suppl 5): 13–23
143 Ballou LR, Botting RM, Goorha S et al (2000) Nociception in cyclooxygenase isozyme-deficient mice. *Proc Natl Acad Sci USA* 97: 10272–10276
144 Breder CD, DeWitt D, Kraig RP (1995) Characterization of inducible cyclooxygenase in rat brain. *J Comp Neurol* 355: 296–315
145 Breder CD, Smith WL, Raz A et al (1992) Distribution and characterization of cyclooxygenase immunoreactivity in the ovine brain. *J Comp Neurol* 322: 409–438
146 Yamagata K, Andreasson KI, Kaufmann WE et al (1993) Expression of a mitogen-inducible cyclooxygenase in brain neurons: regulation by synaptic activity and glucocorticoids. *Neuron* 11: 371–386
147 Matsumura K, Cao C, Ozaki M et al (1998) Brain endothelial cells express cyclooxygenase-2 during lipopolysaccharide-induced fever: light and electron microscopic immunocytochemical studies. *J Neurosci* 18: 6279–6289
148 Ebersberger A, Averbeck B, Messlinger K, Reeh PW (1999) Release of substance P, calcitonin gene-related peptide and prostaglandin E2 from rat dura mater encephali following electrical and chemical stimulation *in vitro*. *Neuroscience* 89: 901–907
149 Sugimoto Y, Shigemoto R, Namba T et al (1994) Distribution of the messenger RNA for

the prostaglandin E receptor subtype EP3 in the mouse nervous system. *Neuroscience* 62: 919–928

150 Hashimoto M, Yamamoto Y, Takagi H (1997) Effects of KB-2796 on plasma extravasation following antidromic trigeminal stimulation in the rat. *Res Commun Mol Pathol Pharmacol* 97: 79–94

151 Iversen HK, Olesen J (1996) Headache induced by a nitric oxide donor (nitroglycerin) responds to sumatriptan. A human model for development of migraine drugs. *Cephalalgia* 16: 412–418

152 Thomsen LL, Olesen J (1998) Nitric oxide theory of migraine. *Clin Neurosci* 5: 28–33

153 Thomsen LL, Olesen J (1997) A pivotal role of nitric oxide in migraine pain. *Ann NY Acad Sci* 835: 363–372

154 Salvemini D, Currie MG, Mollace V (1996) Nitric oxide-mediated cyclooxygenase activation. A key event in the antiplatelet effects of nitrovasodilators. *J Clin Invest* 97: 2562–2568

155 Reuter U, Olesen IJ, Sanchez del Rio M et al (2000) Nitroglycerin infusion induces nitric oxide synthase type II expression in rat dura mater (abstract). *Cephalalgia* 20: 281–282

156 Caterina MJ, Schumacher MA, Tominaga M et al (1997) The capsaicin receptor: a heat-activated ion channel in the pain pathway. *Nature* 389: 816–824

157 Tominaga M, Caterina MJ, Malmberg AB et al (1998) The cloned capsaicin receptor integrates multiple pain-producing stimuli. *Neuron* 21: 531–543

158 Guo A, Vulchanova L, Wang J et al (1999) Immunocytochemical localization of the vanilloid receptor 1 (VR1): relationship to neuropeptides, the P2X3 purinoceptor and IB4 binding sites. *Eur J Neurosci* 11: 946–958

159 Schepelmann K, Ebersberger A, Pawlak M et al (1999) Response properties of trigeminal brain stem neurons with input from dura mater encephali in the rat. *Neuroscience* 90: 543–554

Mast cell involvement in neurogenic inflammation

Theoharis C. Theoharides and Kristiana Kandere

Tufts University School of Medicine, New England Medical Center, 136 Harrison Avenue, Boston, MA 0211, USA

Localization and function of mast cells

Mast cells derive from a distinct precursor in the bone marrow [1] and migrate into most tissues, where they acquire distinct characteristics in response to different micro-environmental influences, such as stem cell factor, nerve growth factor, or the cytokines, interleukin (IL)-3, -4 and -6 [2]. Mast cells are responsible for allergic reactions but mounting evidence indicates that they also participate in inflammation [3, 4] and homeostasis [5]. They are present in the meninges [6–11], especially the dura mater, which contains a large proportion of the total intracranial histamine [12]. The mast cells in the dura resemble those in connective tissue since they stain metachromatically with toluidine blue [13] and immunohistochemically for rat mast-cell protease I [14]. Connective-tissue mast cells and mucosal mast cells vary considerably in their staining characteristics as well as in their cytokine content [2, 15]. The mast cells in the rat dura mater, staining violet with toluidine blue, are often seen next to blood vessels (Fig. 1A). This close association with blood vessels can be appreciated better with electron microscopy, showing mast cells "embracing" the endothelial cells that make up the blood vessel wall (Fig. 1B).

The meninges are innervated by catecholaminergic, cholinergic and peptidergic nerve fibers [16–18]; the dura in particular receives rich peptidergic [19] as well as serotonergic innervation [20]. Mast cells in the dura mater are often found in close association with nerve fibers [20–22], of which many are positive for substance P and calcitonin gene-related peptide [20, 23]. Close anatomical localization of mast cells with neurons [24, 25] has also been noted in the hypothalamus [13], dermis [26], bladder [27–29] and mesentery [27, 28, 30, 31]. Dural mast cells can be appreciated adjacent to neuronal processes with apparent points of contact between them (Fig. 2A); such neurons are often immunocytochemically positive for substance P (Fig. 2B). Close association of dura mast cells with nerve fibers is also evident with scanning electron microscopy, where characteristically granular mast cells appear to encircle (Fig. 3A) or touch (Fig. 3B) neuronal processes.

Migraine: A Neuroinflammatory Disease?, edited by Egilius L.H. Spierings and Margarita Sánchez del Río

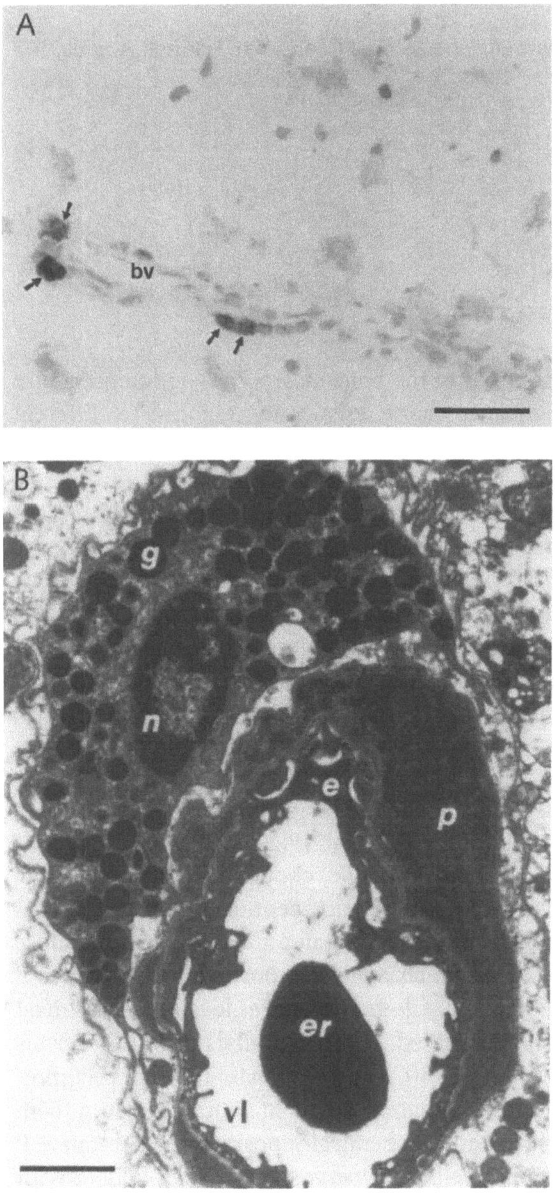

Figure 1

Light photomicrograph of dura mater stained with toluidine blue. (A) Note the numerous mast cells (black arrows) around a blood vessel (bv); Bar = 40 µm; (B) Photomicrograph of a dura perivascular mast cell examined with electron microscopy: er, erythrocyte; p, pericyte; vl, vessel lumen; n, nucleus; g, granule; bar, 4 µm.

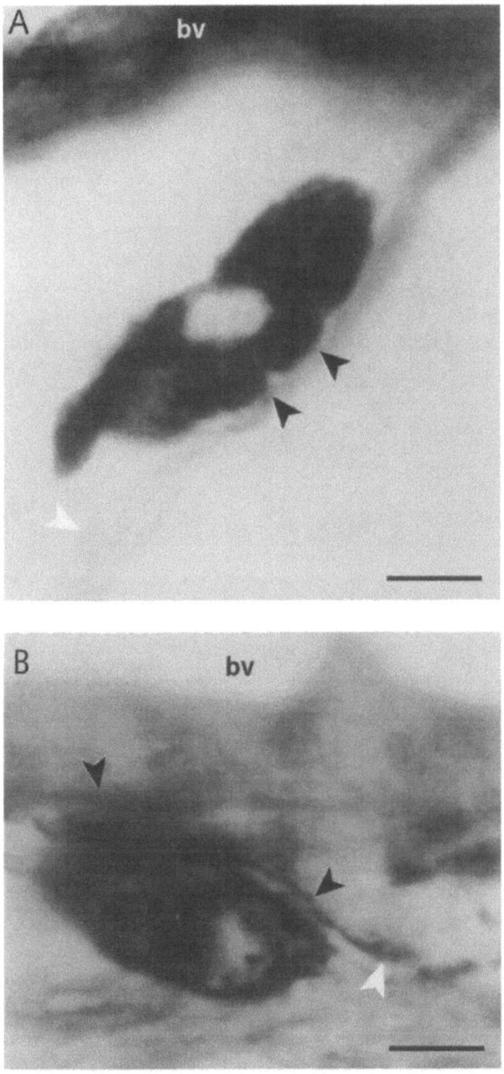

Figure 2
Light photomicrographs of dura mater stained (A) with toluidine blue, showing a mast cell located close to a neuronal process (white arrowheads); black arrowheads indicate points of apparent contact between nerve fiber and mast cell; white arrowhead denotes the nerve fiber. Bar = 4 μm. (B) stained immunohistochemically for substance P and counterstained with toluidine blue, showing a dura mast cell (violet) adjacent to a blood vessel (bv) and to substance P-reactive neuronal processes (white arrowhead); black arrowheads indicate points of apparent contact between nerve fiber and mast cell. Bar = 5 μm

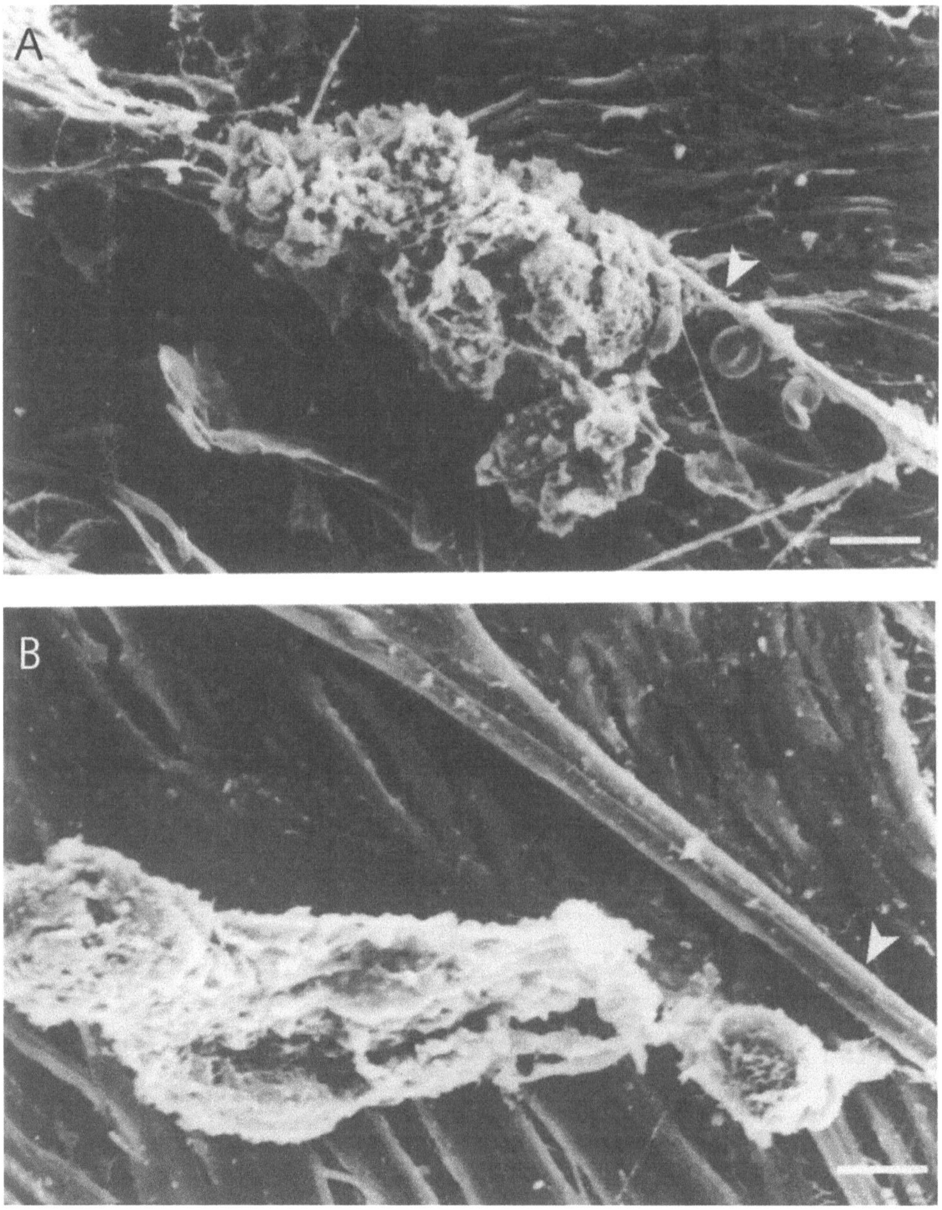

Figure 3
Scanning electron micrographs of dura mast cells (A) wrapped around a nerve fiber or (B) touching a nerve fiber (white arrowhead). Bars = 8 μm.

Neurogenic inflammation

It is now recognized that stress or some other precipitating event can lead to a wave of neuronal dysfunction in the brain ("spreading depression") [32]. Neurogenic inflammation in the dura mater was documented following stimulation of the trigeminal ganglion [33] and has long been considered important in the pathogenesis of the migraine headache [34]. The term "neurogenic inflammation" usually implies activation of some process through neurokinin-1 receptors, the prime target for substance P [34]. However, it has also been used to mean involvement of a variety of other chemicals, with or without the involvement of substance P. Chemical activation of dural sensory neurons recently emerged as a possible trigger of headache [35]. Many such neurosensitizing molecules could derive from dural mast cells and include histamine [36], vasoactive intestinal polypeptide [37], nitric oxide [38], IL-6 and tumor necrosis factor (TNF) [2]. Other cytokines released from mast cells could lead to expression of intercellular adhesion molecules and to the development of inflammation [39, 40]. Vascular permeability dependent on mast cell activation could further explain the pathogenesis of the migraine headache [41], since the throbbing pain [42] is associated with intense vasodilation [43], in which nitric oxide is thought to play a role [44].

Dural mast cells are activated by substance P and calcitonin gene-related peptide [45, 46], as well as by antidromic trigeminal [21], cervical [47], or sphenopalatine [48] ganglion stimulation, leading to increased vascular permeability in rodents [49]. Such findings have raised speculations that interactions between mast cells and neurons [50] may also be involved in neuroinflammatory [3, 51–53] and central nervous system disorders [54–56].

Stress and migraine

It is well known that the migraine headache can be precipitated by emotional or physical stress [57]. Stress triggers the release of corticotropin-releasing hormone, which, in turn, activates the hypothalamic-pituitary-adrenal axis [58]. We previously showed that acute stress, induced by exposing rats to brief periods of immobilization, led to activation of mast cells in the dura mater and increased the level of rat mast-cell protease I in the cerebrospinal fluid [59]. This process was inhibited by neutralizing serum against corticotropin-releasing hormone and was prevented in animals depleted of sensory neuropeptides [59]. Moreover, the antagonist of the receptor for corticotropin-releasing hormone, antalarmin, also inhibited the activation of dural mast cells by acute stress [60]. Immobilization stress for 30 minutes [60] increased mast cell activation in the rat dura mater from 27.9 ± 9.7% to 52.3 ± 7.9%. Pre-treatment intraperitoneally either with azatadine, a histamine/serotonin-receptor antagonist, or the histamine-3-receptor agonist, N^{α}-methylhistamine,

Table 1 - Effect of azatadine and N^{α}-methylhistamine on stress-induced activation of mast cells in the dura mater

	mast cell activation (% total)
Control	27.9 ± 9.7
Stress	52.3 ± 7.9
Azatadine	30.1 ± 6.7
N^{α}-methylhistamine	29.9 ± 6.7
Azatadine + stress	35.0 ± 9.7*
N^{α}-methylhistamine + stress	32.3 ± 5.8*

*$p < 0.05$

30 min prior to stress, inhibited ($p < 0.05$) the activation of mast cells to 35.0 ± 9.2% and 32.3 ± 5.8%, respectively (Tab. 1). We subsequently showed that corticotropin-releasing hormone and its structural analog, urocortin, stimulated mast cell activation and increased vascular permeability directly in rodent skin [61]. This action was also inhibited by antalarmin and was absent in mice deficient in mast cells [61]. We hypothesized that acute stress, through releasing corticotropin-releasing hormone or urocortin, could activate the trigeminal sensory nerve fibers to release neuropeptides, which could then activate the dural mast cells. This premise is supported by the fact that expression of the receptor for corticotropin-releasing hormone was increased on the trigeminal nucleus caudalis, in response to acute immobilization stress [62]. The concept of neurogenic inflammation [34] should, therefore, be broadened to include corticotropin-releasing hormone and related molecules, acting at a step prior to, or in the absence of, the sensory neuropeptides.

Immobilization of mice for 30 min led to significant extravasation of 99-technetium-gluceptate in the dura mater (n = 52), reported as percentage change from control [63]. This effect was totally absent in mice deficient in mast cells [64] subjected to immobilization stress for the same period of time. The change was statistically significant ($p < 0.001$) and indicated that the increase in vascular permeability in the dura mater was mast cell-dependent [64].

Acute immobilization stress also increased the extravasation of 99-technetium gluceptate in the rat diencephalon, where mast cells are present, but not in the cerebral cortex, where they are absent [63]; this action was also mast cell-dependent as it was blocked by the "mast cell stabilizer", disodium cromoglycate [63]. However, unlike acute stress that increases vascular permeability in both the dura mater and diencephalon [63], neurogenic plasma-protein extravasation due to stimulation of the trigeminal nerve was previously reported in the dura mater only and not in the brain itself [33].

Our results indicate that stress-induced activation of dural mast cells can be prevented by pre-treatment with the mixed histamine/serotonin-receptor antagonist, azatadine, as well as by the histamine-3-receptor agonist, N^{α}-methylhistamine. These findings are particularly relevant to migraine because azatadine is structurally similar to cyproheptadine, which has been used for the preventive treatment of migraine [65]. Histamine-3-receptor agonists have previously been shown to inhibit the neuronal release of histamine [66] and many other neurotransmitters [67]. They also inhibit the release of histamine from rat brain slices in a perfusion chamber [68] and block neurogenic plasma-protein extravasation in the rat dura mater, induced by stimulation of the trigeminal nerve [69]. These effects may be mediated by histamine-3-receptor agonists acting on C-fibers making contact with mast cells [70]. In fact, a histamine-3-receptor agonist pro-compound was recently shown to have both anti-inflammatory and analgesic effects [71].

Mast cell involvement in migraine

Even though there is no reliable animal model for migraine [72], increasing evidence indicates that stimulation of the trigeminal nerve leads to vasodilation and neurogenic inflammation [73]. Activation of dural mast cells in response to antidromic trigeminal [21], cervical [47] or parasympathetic sphenopalatine [48] ganglion stimulation could lead to release of numerous vasodilatory and pro-inflammatory molecules [2,3] relevant to migraine [41]. Moreover, long-term superior cervical sympathectomy leads to mast cell hyperplasia and elevations in the content of histamine and serotonin in the rat dura mater [74]. Medications used for the abortive treatment of migraine inhibit mast cell activation in the dura mater and increase vascular permeability, induced by antidromic stimulation of the trigeminal nerve [75]. Dural mast cell involvement in the pathogenesis of migraine is also supported by findings showing degranulated mast cells in biopsies from temporal arteries only on the painful side of patients with cluster headache [76–78]. These patients are thought to have some impairment of the sympathetic nervous system, leading to local mastocytosis [79]. In addition, patients with cluster headache [80] or migraine [81] have elevated serum levels of histamine and administration of histamine leads to intense headache [82]. We recently showed that migraine headaches in children are associated with increased levels of the unique mast cell enzyme, tryptase, in the urine, which decreases together with the duration and intensity of the headaches when the patients are taught to relax [83].

Mast cells in connective tissues can be readily activated by neuropeptides [84], such as substance P [46, 85–87], while mucosal mast cells are weakly responsive only to substance P [88–90]. Release of substance P or calcitonin gene-related peptide from sensory nerve fibers could, therefore, stimulate activation of mast cells in the dura mater [3, 90], as shown previously [45, 46, 60]. Functional consequences of the interactions of mast cells and nerve fibers in the dura mater were reported recently [22, 60]. Activation of mast cells by substance P [46, 86, 87] or other neuropeptides could lead to granulocyte infiltration and neurogenic inflammation [49, 91, 92], possibly through release of TNF, which can then induce expression of endothelial adhesion molecules [93]. These include histamine [80, 81], vasoactive intestinal polypeptide [37, 94] and nitric oxide [38, 44], which could also sensitize sensory nerve fibers in the dura mater and lead to headache [35]. In order to ascertain whether dural vasodilation induced by mast cell mediators could be detected *in vivo*, histamine (0.1 mmol/L) was administered intravenously (0.5 ml) to anesthetized rats. Blood flow and blood volume were assessed by fiberoptic probes placed on the forehead with laser beam excitation at wavelengths to measure oxygenated and deoxygenated hemoglobin. Histamine rapidly decreased cerebral blood flow with a corresponding increase in blood volume, as shown in Figure 4. Nevertheless, emphasis on a single molecule only is very likely to be too simplistic. A case in point is nitric oxide, which has been considered critical in the pathogenesis of the migraine headache [44, 95]. However, nitric oxide appears to *inhibit* activation of mast cells [96] and inflammation dependent on it [97, 98]. In fact, stimulation of nitric oxide synthase by N^G-nitro-L-arginine *increases* vascular permeability induced by administration of corticotropin-releasing hormone [61].

Pro-inflammatory effects of corticotropin-releasing hormone

Corticotropin-releasing hormone is a 41-amino-acid peptide, which regulates the hypothalamic-pituitary-adrenal axis and mediates the stress response [99] through specific receptors [100, 101]. Administration of the hormone causes peripheral vasodilation seen as flushing and itching, symptoms consistent with mast cell activation [102, 103]. The hormone is also secreted outside the central nervous system where it has pro-inflammatory actions [58, 104]. Corticotropin-releasing hormone-like immunoreactivity is also found in the dorsal horn of the spinal cord and the dorsal root ganglia [30, 105, 106], as well as in sympathetic nerve cell bodies and sympathetic ganglia [105, 107, 108]. Antidromic or orthodromic nerve activation could, therefore, lead to the release of corticotropin-releasing hormone and subsequent activation of mast cells. A direct *in vivo* action of corticotropin-releasing hormone on dural blood vessels cannot be excluded because receptors for this hormone have been identified on brain arterioles [109] and human umbilical veins [110, 111]. Both immunoreactive corticotropin-releasing hormone and messenger RNA for the hormone have also been demonstrated in leukocytes, while corticotropin-releasing hormone also stimulates lymphocyte proliferation, increases the expression of IL-2-receptor on T lymphocytes, and is chemotactic for mononuclear leukocytes [112]. Migraine headaches worsen with physical or emotional stress [57, 72]. Acute immobilization stress induces mast cell activation in the dura mater, which is blocked by

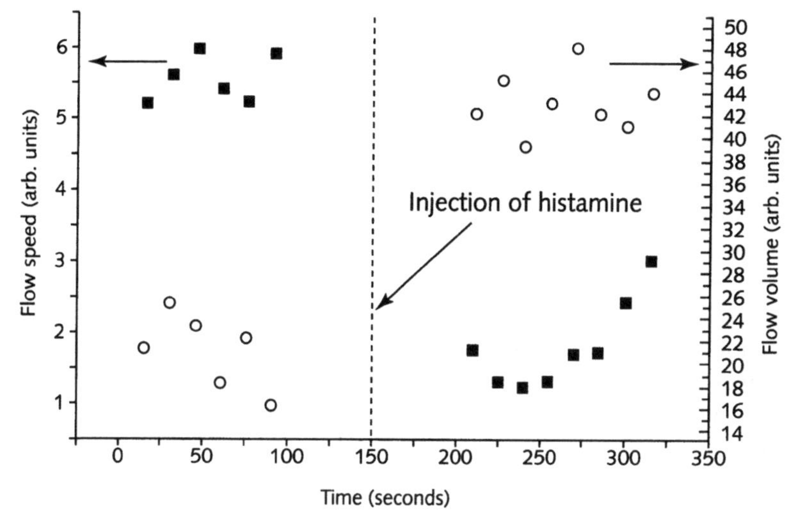

Figure 4
Dura mater vascular blood flow and volume in response to an intravenous injection of 0.5 ml of 0.1 mmol/L of histamine recorded by fiberoptic probes positioned on the forehead of a rat following excitation by a laser beam.

pre-treatment neonatally with capsaicin, indicating the involvement of sensory neuropeptides *in vivo* [59, 61]. It is also inhibited by pre-treatment with a neutralizing antiserum for corticotropin-releasing hormone as well as by the non-peptide antagonist of corticotropin-releasing hormone-receptor, antalarmin [60]. Corticotropin-releasing hormone has been shown to increase vascular permeability in rodent skin, through an action involving receptors for corticotropin-releasing hormone and mast cells [61]. Moreover, leukemic mast cells have been shown to express these receptors [61, 113].

Such findings indicate that interactions between mast cells and neurons [50] may be involved in neuroinflammatory [3, 51–53] and central nervous system disorders [54–56]. This premise is further supported by the fact that patients with systemic mastocytosis exhibit frequent headaches [114, 115] and by the recent demonstration that the unique mast cell enzyme, tryptase, is elevated in the urine of children with migraine and is reduced following training in self-relaxation [83]. Even the abortive antimigraine medication, sumatriptan, has been shown to also act presynaptically to block mast cell activation in the dura mater [116] and neurogenic plasma-protein extravasation [34, 117]; similarly, the newer medication, rizatriptan, blocked cerebral vasodilation and plasma-protein extravasation [118]. This presynaptic mechanism of action now appears to be prominent in the action of triptans [119].

Therapeutic approaches to migraine prevention

There is no truly effective preventive treatment for migraine [120, 121]. Even though substance P is involved in pain transmission [122] and inflammation [123], as also shown with the use of neurokinin-1-receptor knockout mice, neurokinin-1-receptor antagonists have failed to induce effective analgesia in man [124, 125]. In one report, two double blind, placebo-controlled clinical trials using CP-122288, a potent inhibitor of neurogenic inflammation in animals, failed to show an *abortive* antimigraine effect [126]. The authors concluded that neurogenic plasma-protein extravasation is unlikely to play a crucial role in the pathogenesis of migraine headache [126].

Increasing evidence indicates that local neurogenic inflammation, possibly in response to spreading depression, and not necessarily simply vasodilation or plasma-protein extravasation, may be involved in sensory nerve sensitization and migraine generation [32]. Our results indicate that dural mast cells may be critical in this process through the action of corticotropin-releasing hormone. The term "neurogenic" may, therefore, have to be broadened to include corticotropin-releasing hormone, which acts at a step *before* the action of sensory neuropeptides, possibly to induce their release. This is supported by the finding that acute immobilization stress induces corticotropin-releasing hormone receptor expression on the principal sensory nucleus of the trigeminal nerve, the trigeminal nucleus caudalis [62].

Stress hormones such as corticotropin-releasing hormone can orthodromically activate trigeminal nerve fibers to release neuropeptides. Presynaptic inhibition at this stage could be achieved either by antagonists of the corticotropin-releasing hormone receptor, which block neuronal excitation by stress, or by histamine-3-receptor agonists, which inhibit the release of serotonin or neuropeptides. Corticotropin-releasing hormone and any neuropeptides released could activate mast cells in the dura mater to secrete vasoactive nociceptive and pro-inflammatory molecules. Serotonin, which is a stronger vasodilator in rodents than histamine, could then cause vasodilation and sensitization of peripheral nerve fibers. This process could be inhibited either by histamine-3-receptor agonists, which inhibit the release of serotonin, or histamine/serotonin-receptor antagonists, such as azatadine.

Acknowledgements
The work presented in this chapter was supported in part by grant No. N538326 from the USA National Institutes of Health and Theta Biomedical Consulting and Development Co., Inc. (Brookline, MA, USA).

References

1 Rodewald H-R, Dessing M, Dvorak AM, Galli SJ (1996) Identification of a committed precursor for the mast cell lineage. *Science* 271: 818–822

2 Galli SJ (1993) New concepts about the mast cell. *N Engl J Med* 328: 257–265

3 Theoharides TC (1996) Mast cell: a neuroimmunoendocrine master player. *Int J Tissue React* 18: 1–21

4 McNeil HP, Gotis-Graham I (2000) Human mast cell subsets-distinct functions in inflammation. *Inflamm Res* 49: 3–7

5 Gurish MF, Austen KF (2001) The diverse roles of mast cells. *J Exp Med* 194: F1–F5

6 Dropp JJ (1979) Mast cells in the human brain. *Acta Anat* 105: 505–513

7 Edvinsson L, Cervos-Navarro J, Larsson LI et al (1977) Regional distribution of mast cells containing histamine, dopamine or 5-hydroxytryptamine in the mammalian brain. *Neurology* 27: 878–884

8 Edvinsson L, Owman C, Sjöberg NO (1976) Autonomic nerves, mast cells and amine receptors in human brain vessels. A histochemical and pharmacological study. *Brain Res* 115: 377–393

9 Pollard H, Bischoff S, Llorens-Cortes C, Schwartz JC (1976) Histidine decarboxylase and histamine in discrete nuclei of rat hypothalamus and the evidence for mast cells in the median eminence. *Brain Res* 118: 509–513

10 Ibrahim MZ (1974) The mast cells of the mammalian central nervous system. Part I. Morphology, distribution and histochemistry. *J Neurol Sci* 21: 431–478

11 Olsson Y (1968) Mast cells in the nervous system. *Int Rev Cytol* 24: 27–70

12 Orr EL, Pace KR (1984) The significance of dura mast cells as a source of histamine in the mouse brain. *J Neurochem* 42: 727–732

13 Pang X, Letourneau R, Rozniecki JJ et al (1996) Definitive characterization of rat hypothalamic mast cells. *Neuroscience* 73: 889–902

14 Dimitriadou V, Rouleau A, Tuong T et al (1996) Rat cerebral mast cells undergo phenotype changes during development. *Dev Brain Res* 97: 29–41

15 Bradding P, Okayama Y, Howarth PH et al (1995) Heterogeneity of human mast cells based on cytokine content. *J Immunol* 155: 297–307

16 Fricke B, Von During M, Andres KH (1997) Topography and immunocytochemical characterization of nerve fibers in the leptomeningeal compartments of the rat. A light- and electron-microscopical study. *Cell Tissue Res* 287: 11–22

17 Von During M, Fricke B, Dahlmann A (1995) Topography and distribution of nerve fibers in the posterior longitudinal ligament of the rat: an immunocytochemical and electron-microscopical study. *Cell Tissue Res* 281: 325–338

18 Von During M, Andres KH (1991) Sensory nerve fiber terminals in the arachnoid granulations of non-human primates. *Neurosci Lett* 127: 121–124

19 Von During M, Bauersachs M, Bohmer B et al (1990) Neuropeptide Y- and substance P-like immunoreactive nerve fibers in the rat dura mater encephali. *Anat Embryol (Berl)* 182: 363–373

20 Keller JT, Marfurt CF (1991) Peptidergic and serotoninergic innervation of the rat dura mater. *J Comp Neurol* 309: 515–534

21 Dimitriadou V, Buzzi MG, Moskowitz MA, Theoharides TC (1991) Trigeminal sensory fiber stimulation induces morphologic changes reflecting secretion in rat dura mast cells. *Neuroscience* 44: 97–112

22 Dimitriadou V, Rouleau A, Trung Tuong MD et al (1997) Functional relationships between sensory nerve fibers and mast cells of dura mater in normal and inflammatory conditions. *Neuroscience* 77: 829–839

23 Lambracht-Hall M, Dimitriadou V, Theoharides TC (1990) Migration of mast cells in the developing rat brain. *Dev Brain Res* 56: 151–159

24 Dimitriadou V, Aubineau P, Taxi J, Seylaz J (1987) Ultrastructural evidence for a functional unit between nerve fibers and type II cerebral mast cells in the cerebral vascular wall. *Neuroscience* 22: 621–630

25 Newson B, Dahlström A, Enerbäck L, Ahlman H (1983) Suggestive evidence for a direct innervation of mucosal mast cells. *Neuroscience* 10: 565–570

26 Naukkarinen A, Harvima I, Paukkonen K et al (1993) Immunohistochemical analysis of sensory nerves and neuropeptides and their contacts with mast cells in developing and mature psoriatic lesions. *Arch Dermatol Res* 285: 341–346

27 Bienenstock J, Tomioka M, Matsuda H et al (1987) The role of mast cells in inflammatory processes: evidence for nerve mast cell interactions. *Int Arch Allergy Appl Immunol* 82: 238–243

28 Stead RH, Tomioka M, Quinonez G et al (1987) Intestinal mucosal mast cells in normal and nematode-infected rat intestines are in intimate contact with peptidergic nerves. *Proc Natl Acad Sci USA* 84: 2975–2979

29 Letourneau R, Pang X, Sant GR, Theoharides TC (1996) Intragranular activation of bladder mast cells and their association with nerve processes in interstitial cystitis. *Br J Urol* 77: 41–54

30 Skofitsch G, Savitt JM, Jacobowitz DM (1985) Suggestive evidence for a functional unit between mast cells and substance P fibers in the rat diaphragm and mesentery. *Histochemistry* 82: 5–8

31 Stead RH, Dixon MF, Bramwell NH et al (1989) Mast cells are closely apposed to nerves in the human gastrointestinal mucosa. *Gastroenterology* 97: 575–585

32 Aurora SK, Welch KM (2000) Migraine: imaging the aura. *Curr Opin Neurol* 13: 273–276

33 Markowitz S, Saito K, Moskowitz MA (1987) Neurogenically mediated leakage of plasma protein occurs from blood vessels in dura mater but not brain. *J Neurosci* 7: 4129–4136

34 Moskowitz MA (1993) Neurogenic inflammation in the pathophysiology and treatment of migraine. *Neurology* 43: S16–S20

35 Strassman AM, Raymond SA, Burstein R (1996) Sensitization of meningeal sensory neurons and the origin of headaches. *Nature* 384: 560–564

36 LaMerte FP, Merchant JA, Casale TB (1988) Acute systemic reactions to carbonless copy paper associate with histamine release. *J Am Med Assoc* 260: 242–243

37 Lutz E, Chan W, Track NS et al (1978) Release of vasoactive intestinal polypeptide in mast cells by histamine liberators. *Nature* 275: 661–662

38 Mannaioni PF, Masini E, Pistelli A et al (1991) Mast cells as a source of superoxide anions and nitric oxide-like factor: relevance to histamine release. *Int J Tissue React* 13: 271–278

39 Klein LM, Lavker RM, Matis WL, Murphy GF (1989) Degranulation of human mast cells induces an endothelial antigen central to leukocyte adhesion. *Proc Natl Acad Sci USA* 86: 8972–8976

40 Suzuki H, Miura S, Liu YY et al (1995) Substance P induces degranulation of mast cells and leukocyte adhesion to venular endothelium. *Peptides* 16: 1447–1452

41 Theoharides TC (1983) Mast cells and migraines. *Perspect Biol Med* 26: 672–675

42 Kudrow L, Kudrow DB, Sandweiss JH (1995) Rapid and sustained relief of migraine attacks with intranasal lidocaine: preliminary findings. *Headache* 35: 79–82

43 Olesen J, Friberg L, Olesen TS et al (1990) Timing and topography of cerebral blood flow, aura and headache during migraine attacks. *Ann Neurol* 28: 791–798

44 Olesen J, Thomsen LL, Iversen H (1994) Nitric oxide is a key molecule in migraine and other vascular headaches. *Trends Pharmacol Sci* 15: 149–153

45 Dimitriadou V, Lambracht-Hall M, Theoharides TC (1990) Can inhibition of carbachol or substance P induced selective release from mast cells be associated with analgesia (abstract). *Pain* 5 (Suppl): 514

46 Ottosson A, Edvinsson L (1997) Release of histamine from dural mast cells by substance P and calcitonin gene-related peptide. *Cephalalgia* 17: 166–174

47 Keller JT, Dimlich RV, Zuccarello M et al (1991) Influence of the sympathetic nervous system as well as trigeminal sensory fibers on rat dural mast cells. *Cephalalgia* 11: 215–221

48 Delepine L, Aubineau P (1997) Plasma protein extravasation induced in the rat dura mater by stimulation of the parasympathetic sphenopalatine ganglion. *Exp Neurol* 147: 389–400

49 Dimitriadou V, Buzzi MG, Theoharides TC, Moskowitz MA (1992) Ultrastructural evidence for neurogenically mediated changes in blood vessels of the rat dura mater and tongue following antidromic trigeminal stimulation. *Neuroscience* 48: 187–203

50 Williams RM, Bienenstock J, Stead RH (1995) Mast cells: the neuroimmune connection. *Chem Immunol* 61: 208–235

51 Foreman JC (1987) Neuropeptides and the pathogenesis of allergy. *Allergy* 42: 1–11

52 Theoharides TC (1990) Mast cells: the immune gate to the brain. *Life Sci* 46: 607–617

53 Marshall JS, Waserman S (1995) Mast cells and the nerves – potential interactions in the context of chronic disease. *Clin Exp Allergy* 25: 102–110

54 Silver R, Silverman A-J, Vitkovic L, Lederhendler II (1996) Mast cells in the brain: evidence and functional significance. *Trends Neurosci* 19: 25–31

55 Kines KC, Powell HC (1997) Mast cell interactions with the nervous system: relationship to mechanisms of disease. *J Neuropathol Exp Neurol* 56: 627–640
56 Purcell WM, Westgate C, Atterwill CK (1996) Rat brain mast cells: an *in vitro* paradigm for assessing the toxic effects of neurotropic therapeutics. *NeuroToxicol* 17: 845–850
57 Rasmussen BK (1993) Migraine and tension-type headache in a general population: precipitating factors, female hormones, sleep pattern and relation to life style. *Pain* 53: 65–72
58 Chrousos GP (1995) The hypothalamic-pituitary-adrenal axis and immune-mediated inflammation. *N Engl J Med* 332: 1351–1362
59 Theoharides TC, Spanos CP, Pang X et al (1995) Stress-induced intracranial mast cell degranulation. A corticotropin releasing hormone-mediated effect. *Endocrinology* 136: 5745–5750
60 Rozniecki JJ, Dimitriadou V, Lambracht-Hall M et al (1999) Morphological and functional demonstration of rat dura mast cell-neuron interactions *in vitro* and *in vivo*. *Brain Res* 849: 1–15
61 Theoharides TC, Singh L, Boucher W et al (1998) Corticotropin-releasing hormone induces skin mast cell degranulation and increased vascular permeability, a possible explanation for its pro-inflammatory effects. *Endocrinology* 139: 403–413
62 Rivest S, Laflamme N, Nappi RE (1995) Immune challenge and immobilization stress induce transcription of the gene encoding the CRF receptor in selective nuclei of the rat hypothalamus. *J Neurosci* 15: 2680–2695
63 Esposito P, Gheorghe D, Kandere K et al (2001) Acute stress increases permeability of the blood-brain-barrier through activation of brain mast cells. *Brain Res* 888: 117–127
64 Kandere K, Dimitriadou V, Gheorge D et al (2001) Acute stress increases vascular extravasation of ^{99}techetium-gluceptate in the dura mater but not in mast cell deficient, neurokinin-1 or substance P knock-out mice. *J Clin Invest; in press*
65 Diamond S (1991) Migraine headaches. *Med Clin North Am* 75: 545–566
66 Van der Werf JF, Timmerman H (1989) The histamine H_3 receptor: a general presynaptic histaminergic regulatory system? *TIPS* 10: 159–162
67 Leurs R, Blandina P, Tedford C, Timmerman H (1998) Therapeutic potential of histamine H_3 receptor agonists and antagonists. *TIPS* 19: 177–183
68 Rozniecki JJ, Letourneau R, Sugiultzoglu M et al (1999) Differential effect of histamine-3 receptor active agents on brain but not peritoneal mast cell activation. *J Pharmacol Exp Ther* 290: 1–9
69 Matsubara T, Moskowitz MA, Huang Z (1992) UK-14,304 R(–)-α-methyl-histamine and SMS 201-995 block plasma protein leakage within dura mater by prejunctional mechanisms. *Eur J Pharmacol* 224: 145–150
70 Dimitriadou V, Rouleau A, Dam Trung Tuong M et al (1994) Functional relationship between mast cells and C-sensitive nerve fibers evidenced by histamine H_3-receptor modulation in rat lung and spleen. *Clin Sci* 87: 151–163
71 Rouleau A, Stark H, Schunack W, Schwartz J-C (2000) Anti-inflammatory and antinoci-

ceptive properties of BP 2-94, a histamine H_3-receptor agonist prodrug. *J Pharmacol Exp Ther* 295: 219–225
72 Edvinsson L (1995) Experimental headache models in animals and humans. *TIPS* 16: 5–9
73 Moskowitz MA (1992) Neurogenic *versus* vascular mechanisms of sumatriptan and ergot alkaloids in migraine. *TIPS* 13: 307–311
74 Bergerot A, Reynier-Rebuffel AM, Callebert J, Aubineau P (2000) Long-term superior cervical sympathectomy induces mast cell hyperplasia and increases histamine and serotonin content in the rat dura mater. *Neuroscience* 96: 205–213
75 Buzzi MG, Dimitriadou V, Theoharides TC, Moskowitz MA (1992) 5-Hydroxytryptamine receptor agonists for the abortive treatment of vascular headaches block mast cell, endothelial and platelet activation within the rat dura mater after trigeminal stimulation. *Brain Res* 583: 137–149
76 Liberski PP, Prusinski A (1982) Further observations on the mast cells over the painful region in cluster headache patients. *Headache* 22: 115–117
77 Liberski PP, Mirecka B (1984) Mast cells in cluster headache: ultrastructure, release pattern and possible pathogenetic significance. *Cephalalgia* 4: 101–106
78 Dimitriadou V, Henry P, Brochet B et al (1990) Cluster headache: ultrastructural evidence for mast cell degranulation and interaction with nerve fibers in the human temporal artery. *Cephalalgia* 10: 221–228
79 Russell AL, McCarty MF (2000) Glucosamine for migraine prophylaxis. *Med Hypotheses* 55: 195–198
80 Anthony M, Lance JW (1971) Histamine and serotonin in cluster headache. *Arch Neurol* 25: 225–231
81 Heatley RV, Denburg JA, Bayer N, Bienenstock J (1982) Increased plasma histamine levels in migraine patients. *Clin Allergy* 12: 145–149
82 Krabbe AA, Olesen J (1980) Headache provocation by continuous intravenous infusion of histamine, clinical results and receptor mechanisms. *Pain* 8: 253–259
83 Olness K, Hall H, Rozniecki JJ et al (1999) Mast cell activation in children with migraine before and after training in self-regulation. *Headache* 39: 101–107
84 Foreman JC (1987) Peptides and neurogenic inflammation. *Brain Res Bull* 43: 386–398
85 Ali H, Leung KBP, Pearce FL et al (1986) Comparison of the histamine-releasing action of substance P on mast cells and basophils from different species and tissues. *Int Arch Allergy Appl Immunol* 79: 413–418
86 Fewtrell CMS, Foreman JC, Jordan CC et al (1982) The effects of substance P on histamine and 5-hydroxytryptamine release in the rat. *J Physiol* 330: 393–411
87 Hakanson R, Leander S, Andersson RG, Horig J (1983) Substance P antagonists release histamine from peritoneal mast cells. *Acta Physiol Scand* 117: 319–320
88 Shanahan F, Denburg JA, Fox J et al (1985) Mast cell heterogeneity: effects of neuroenteric peptides on histamine release. *J Immunol* 135: 1331–1337
89 Lowman MA, Rees PH, Benyon RC, Church MK (1988) Human mast cell heterogeneity: histamine release from mast cells dispersed from skin, lung, adenoids, tonsils, and

colon in response to IgE-dependent and nonimmunologic stimuli. *J Allergy Clin Immunol* 81: 590–597
90 Janiszewski J, Bienenstock J, Blennerhassett MG (1994) Picomolar doses of substance P trigger electrical responses in mast cells without degranulation. *Am J Physiol* 267: C138–C145
91 Matsuda H, Kawakita K, Kiso Y et al (1989) Substance P induces granulocyte infiltration through degranulation of mast cells. *J Immunol* 142: 927–931
92 Glover V, Sandler M (1989) Can the vascular and neurogenic theories of migraine be reconciled? *TIPS* 10: 1–3
93 Walsh LJ, Trinchieri G, Waldorf HA et al (1991) Human dermal mast cells contain and release tumor necrosis factor α, which induces endothelial leukocyte adhesion molecule 1. *Proc Natl Acad Sci USA* 88: 4220–4224
94 Goadsby PJ, Edvinsson L, Ekman R (1990) Vasoactive peptide release in the extracerebral circulation of humans during migraine headache. *Ann Neurol* 28: 183–187
95 Stepien A (1999) The role of nitrogen oxide in migraine attack. *Neurol Neurochir Pol* 33: 435–444
96 Brooks AC, Whelan CJ, Purcell WM (1999) Reactive oxygen species generation and histamine release by activated mast cells: modulation by nitric oxide synthase inhibition. *Br J Pharmacol* 128: 585–590
97 Gaboury JP, Niu X-F, Kubes P (1996) Nitric oxide inhibits numerous features of mast cell-induced inflammation. *Circulation* 93: 318–326
98 Kubes P, Kanwar S, Niu X-F, Gaboury JP (1993) Nitric oxide synthesis inhibition induces leukocyte adhesion *via* superoxide and mast cells. *FASEB J* 7: 1293–1299
99 Sternberg EM, Chrousos GP, Wilder RL, Gold PW (1992) The stress response and the regulation of inflammatory disease. *Ann Intern Med* 117: 854–866
100 Chalmers DT, Lovenberg TW, Grigoriadis DE et al (1996) Corticotropin-releasing factor receptors: from molecular biology to drug design. *TIPS* 17: 166–172
101 Spiess J, Dautzenberg FM, Sydow S et al (1998) Molecular properties of the CRF receptor. *Trends Endocrinol Metab* 9: 140–145
102 Kubler A, Rothacher G, Knappertz VA et al (1994) Intra and extracerebral blood flow changes and flushing after intravenous injection of human corticotropin-releasing hormone. *Clin Invest* 72: 331–336
103 Udelsman R, Harwood JP, Millan MA et al (1986) Functional corticotropin-releasing factor receptors in the primate peripheral nervous system. *Nature* 319: 146–150
104 Karalis K, Sano H, Redwine J et al (1991) Autocrine or paracrine inflammatory actions of corticotropin-releasing hormone *in vivo*. *Science* 254: 421–423
105 Merchenthaler I, Hynes MA, Vingh S et al (1983) Immunocytochemical localization of corticotropin-releasing factor (CRF) in the rat spinal cord. *Brain Res* 275: 373–377
106 Skofitsch G, Zamir N, Helke CJ et al (1985) Corticotropin-releasing factor-like immunoreactivity in sensory ganglia and capsaicin sensitive neurons of the rat central nervous system: colocalization with other neuropeptides. *Peptides* 6: 307–318
107 Suda T, Tomori N, Tozawa F et al (1984) Distribution and characterization of immuno-

reactive corticotropin-releasing factor in human tissues. *J Clin Endocrinol Metab* 59: 861–866
108 Udelsman R, Gallucci WT, Bacher J et al (1986) Hemodynamic effects of corticotropin-releasing hormone in the anesthetized cynomolgus monkey. *Peptides* 7: 465–467
109 Chalmers DT, Lovenberg TW, DeSouza EB (1995) Localization of novel corticotropin-releasing factor receptor (CRF_2) mRNA expression to specific subcortical nuclei in rat brain: comparison with CRF_1 receptor mRNA expression. *J Neurosci* 10: 6340–6350
110 Clifton VL, Read MA, Leitch IM et al (1995) Corticotropin-releasing hormone-induced vasodilatation in the human fetal-placental circulation: involvement of the nitric oxide-cyclic guanosine 3',5'-monophosphate-mediated pathway. *J Clin Endocrinol Metab* 80: 2888–2893
111 Singh LK, Pang X, Alexacos N et al (1999) Acute immobilization stress triggers skin mast cell degranulation *via* corticotropin-releasing hormone, neurotensin and substance P: A link to neurogenic skin disorders. *Brain Behavior Immunity* 13: 225–239
112 Karalis K, Louis JM, Bae D et al (1997) CRH and the immune system. *J Neuroimmunol* 72: 131–136
113 Singh LK, Boucher W, Pang X et al (1999) Potent mast cell degranulation and vascular permeability triggered by urocortin through activation of CRH receptors. *J Pharm Exp Ther* 288: 1349–1356
114 Horan RF, Austen KF (1991) Systemic mastocytosis: retrospective review of a decade's clinical experience at the Brigham and Women's Hospital. *J Invest Dermatol* 96: 5S–14S
115 Longley J, Duffy TP, Kohn S (1995) The mast cell and mast cell disease. *J Am Acad Dermatol* 32: 545–561
116 Buzzi MG, Moskowitz MA (1990) The antimigraine drug, sumatriptan (GR43175), selectively blocks neurogenic plasma extravasation from blood vessels in dura mater. *Br J Pharmacol* 99: 202–206
117 Buzzi MG, Moskowitz MA (1991) Evidence for $5\text{-}HT_{1B/1D}$ receptors mediating the antimigraine effect of sumatriptan and dihydroergotamine. *Cephalalgia* 11: 165–168
118 Williamson DJ, Shepheard SL, Hill RG, Hargreaves RJ (1997) The novel anti-migraine agent rizatriptan inhibits neurogenic dural vasodilation and extravasation. *Eur J Pharmacol* 328: 61–64
119 Deleu D, Hanssens Y (2000) Current and emerging second-generation triptans in acute migraine therapy:a comparative review. *J Clin Pharmacol* 40: 687–700
120 Bartleson JD (1999) Treatment of migraine headaches. *Mayo Clin Proc* 74: 702–708
121 Deleu D, Hanssens Y, Worthing EA (1998) Symptomatic and prophylactic treatment of migraine:a critical reappraisal. *Clin Neuropharmacol* 21: 267–279
122 De Felipe C, Herrero JF, O'Brien JA et al (1998) Altered nociception, analgesia and aggression in mice lacking the receptor for substance P. *Nature* 392: 394–397
123 Bozic CR, Lu B, Höpken UE et al (1996) Neurogenic amplification of immune complex inflammation. *Science* 273: 1722–1725

124 Hill R (2000) NK_1 (substance P) receptor antagonists – why are they not analgesic in humans? *TIPS* 21: 244–246
125 Goldstein DJ, Wang O, Saper JR et al (1997) Ineffectiveness of neurokinin-1 antagonist in acute migraine: a crossover study. *Cephalalgia* 17: 785–790
126 Roon KI, Olesen J, Diener HC et al (2000) No acute antimigraine efficacy of CP-122,288, a highly potent inhibitor of neurogenic inflammation: results of two randomized, double-blind, placebo-controlled clinical trials. *Ann Neurol* 47: 238–241

Inhibition of neurogenic inflammation in abortive migraine treatment

Egilius L.H. Spierings

Harvard Medical School, Brigham and Women's Hospital, 75 Francis Street, Boston, MA 02115, USA

Introduction

Neurogenic inflammation is mediated through the release of peptides, called neuropeptides, from the sensory nerve fibers involved in nociception. It comprises two processes, plasma-protein extravasation and neurogenic vasodilation, mediated predominantly by substance P and calcitonin gene-related peptide, respectively. Substance P is released from the unmyelinated C fibers and opens up post-capillary venules by acting on neurokinin-1 receptors, allowing the extravasation of plasma proteins. Calcitonin gene-related peptide, on the other hand, is released from the small myelinated Aδ fibers and dilates the arterioles by stimulating receptors of its own.

Inhibition of neurogenic inflammation was suggested, as early as in 1960, as a possible mode of action of ergotamine [1], in addition to the potent vasoconstrictor effect of the medication [2]. The constrictor effect of ergotamine on cranial arteriovenous anastomoses served as the basis for the development of sumatriptan [3]. Sumatriptan was subsequently shown in animals to inhibit neurogenic plasma-protein extravasation selectively in the dura mater [4]. Selective inhibitors of neurogenic inflammation were subsequently developed and tested clinically. The present chapter concerns itself with this group of compounds, which can be divided into those acting prejunctionally and postjunctionally. The prejunctional inhibitors of neurogenic inflammation interfere with the release of neuropeptides from the nerve fibers. The postjunctional inhibitors block the action of the neuropeptides on their targets within the tissues. Both types of compounds inhibit the plasma-protein extravasation following stimulation of the trigeminal ganglion or administration of capsaicin. Capsaicin acts prejunctionally to cause the release of the neuropeptides from the nerve terminals.

So far, eleven selective inhibitors of neurogenic inflammation have been studied clinically (Tab. 1). They have been studied in randomized, double blind, placebo-controlled trials. Patient selection occurred according to the criteria for migraine, as published by the International Headache Society. They were treated during a

Table 1 - Selective inhibitors of neurogenic inflammation

Prejunctionally acting	Postjunctionally acting
• 4991W93	• RPR-100893
• PNU-142633	• LY-303870 (lanepitant)
• LY-334370	• L-758298
• CP-122228	• GR-205171
• RO-470203 (bosetan)	• FK-888
• SB-220453	

Table 2 - Groups of selective prejunctional inhibitors of neurogenic inflammation

- Serotonin-1D-receptor agonists
- Serotonin-1F-receptor agonists
- "5-carboxamide-tryptamine-receptor" agonists
- Endothelin-B-receptor antagonists
- Benzoylamino-benzopyran compounds

migraine headache of moderate or severe intensity. Treatment was considered effective if, after 2 h, it resulted in a decrease in headache intensity to mild or no pain ("2-h headache relief"). All adverse events were considered, whether or not related to the treatment by the investigators ("treatment-emergent adverse events").

Prejunctional inhibitors

Of the 11 selective inhibitors of neurogenic inflammation studied, six act prejunctionally. From a pharmacological perspective, the prejunctional inhibitors of neurogenic inflammation can be divided into five groups, as shown in Table 2.

4991W93 and PNU-142633

4991W93 and PNU-142633 are selective serotonin-1D-receptor agonists. They act prejunctionally on serotonin-1D receptors to inhibit the release of the neuropeptides. They are devoid of cranial vasoconstrictor activity, which is mediated through stimulation of serotonin-1B receptors. 4991W93 was studied in 117 patients who received placebo or 0.125 or 2.5 mg of the compound intravenously [5]. The results

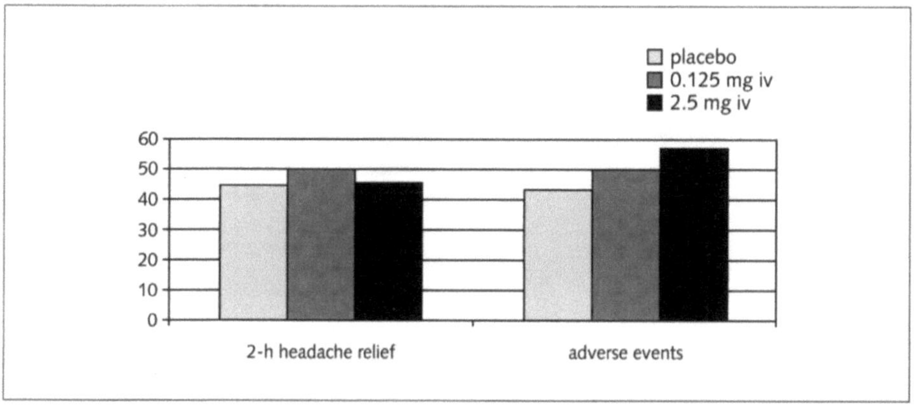

Figure 1
Response to 4991W93, 0.125 and 2.5 mg intravenously, in percentage responders, compared to placebo, in the abortive treatment of moderate or severe migraine headache. Data obtained from [5].

with regard to 2-h headache relief and occurrence of adverse events are shown in Figure 1. There was no benefit, even in those patients who were treated within 2 h of headache onset. PNU-142633 was studied in 69 patients who received either placebo or 50 mg of the compound orally [6]. The results with regard to 2-h headache relief and occurrence of adverse events are shown in Figure 2. The results on headache relief revealed no benefit.

LY-334370

LY-334370 is a selective serotonin-1F-receptor agonist. Of the two identified serotonin-1 receptors (D and F) that inhibit neurogenic inflammation, the F receptor may be the most important one [7]. Like the serotonin-1D-receptor agonists mentioned above, LY-334370 lacks cranial vasoconstrictor activity [8]. It was studied in 131 patients who received placebo or 20, 60, or 200 mg of the compound orally [9]. The results, presented in Figure 3, show efficacy for the higher doses but with considerable adverse events.

CP-122288

CP-122288 is a so-called conformationally restricted sumatriptan analogue [10]. It inhibits plasma-protein extravasation at doses that are devoid of vasoconstrictor

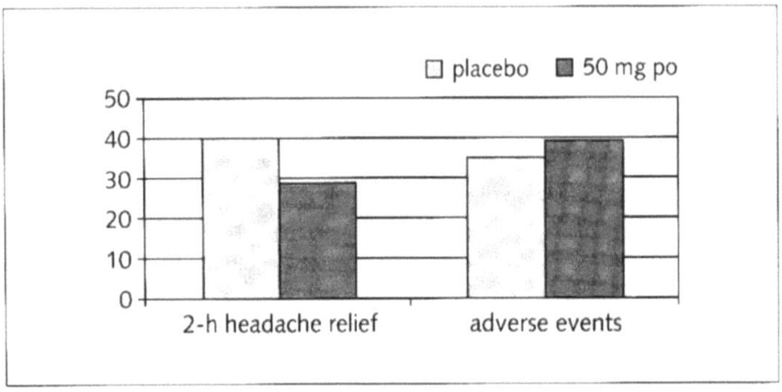

Figure 2
Response to PNU-142633, 50 mg orally, in percentage responders, compared to placebo, in the abortive treatment of moderate or severe migraine headache. Data obtained from [6].

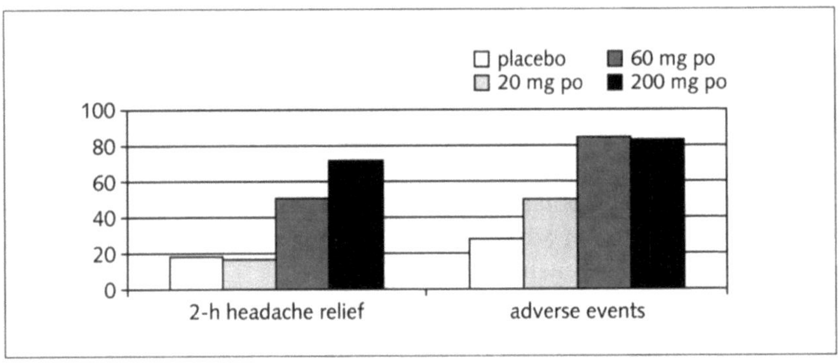

Figure 3
Response to LY-334370, 20, 60, and 200 mg orally, in percentage responders, compared to placebo, in the abortive treatment of moderate or severe migraine headache. Data obtained from [9].

activity [11]. Its potency in inhibiting plasma-protein extravasation does not correlate with its binding affinity at serotonin-1B or -1D receptors [12]. This is similar to 5-carboxamide-tryptamine and both compounds are thought to act on an as yet undefined "extravasation receptor". CP-122288 was studied in 22 patients who received placebo *and* 31.25 µg of the compound intravenously in a crossover design [13]. The results with regard to 2-h headache relief are shown in Figure 4; no information on the occurrence of adverse events is available. The results on headache relief revealed no benefit.

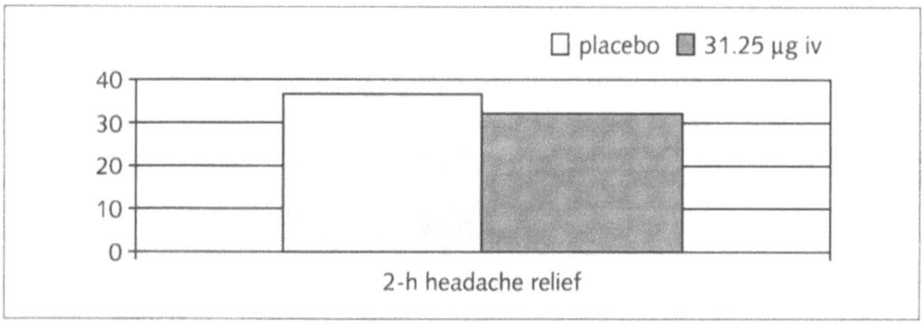

Figure 4
Response to CP-122288, 31.25 µg intravenously, in percentage responders, compared to placebo, in the abortive treatment of moderate or severe migraine headache. Data obtained from [13].

RO-470203 (bosetan)

RO-470203 is an endothelin-receptor antagonist. Endothelins are vasoactive peptides, formed in endothelial cells. They act through two different receptors, referred to as endothelin-A and -B receptors. The endothelin-A receptor mediates the vasoconstrictor action of endothelin-1 in particular. The B-receptor can do the same but also mediates the endothelin-1-induced release of nitric oxide and accompanying vasodilation. Bosetan is an antagonist of both endothelin receptors and lacks vasoconstrictor activity [14]. It inhibits plasma-protein extravasation following stimulation of the trigeminal ganglion but *not* following administration of substance P [15]. This makes it a prejunctional inhibitor of neurogenic inflammation. It was studied in 49 patients who received either placebo or 250 mg of the compound intravenously [16]. The results with regard to 2-h headache relief are shown in Figure 5; no information on the occurrence of adverse events is available. The results on headache relief revealed no benefit.

SB-220453

The last compound in this series, SB-220453, belongs to the group of benzoylamino-benzopyran compounds, which inhibit plasma-protein extravasation through a stereospecific binding site. The neuronal nature of this binding site makes the compounds prejunctional inhibitors of neurogenic inflammation. In addition, they inhibit cortical spreading depression [17]; however, they are devoid of vasoconstrictor activity [18]. SB-220453 was studied in doses of 25, 40, and 80 mg orally. No benefit in terms of migraine relief was observed in comparison to placebo.

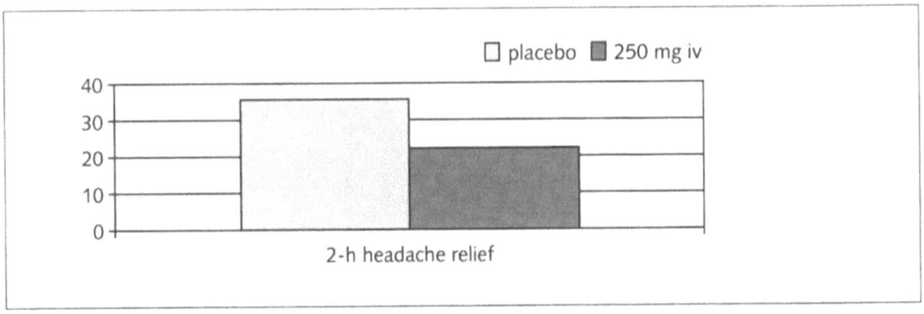

Figure 5
Response to RO-470203, 250 mg intravenously, in percentage responders, compared to placebo, in the abortive treatment of moderate or severe migraine headache. Data obtained from [16].

Postjunctional inhibitors

The five postjunctional inhibitors of neurogenic inflammation studied are neurokinin-1-receptor antagonists; the neurokinin-1 receptor is the prime target for substance P. Neurokinin-1-receptor antagonists are devoid of vasoconstrictor activity but in animals have been shown to have antinociceptive and anti-emetic properties [19]. They can be divided into peptide and non-peptide compounds.

RPR-100893

RPR-100893 is a non-peptide neurokinin-1-receptor antagonist, selective for the human receptor type. It has been shown to inhibit plasma-protein extravasation in the dura mater of the guinea pig, after electrical stimulation of the trigeminal ganglion and administration of capsaicin [20]. It has been studied in the abortive treatment of migraine given orally. No specifics are available with regard to the study, except that its outcome was negative.

LY-303870 (lanepitant)

LY-303870 is also a non-peptide neurokinin-1-receptor antagonist. It has also been shown to inhibit plasma-protein extravasation in the dura mater of the guinea pig, after electrical stimulation of the trigeminal ganglion [21]. It was studied clinically in 40 patients who, in a crossover design, received 30, 80, and 240 mg of the compound orally [22]. The results in terms of 2-h headache relief and

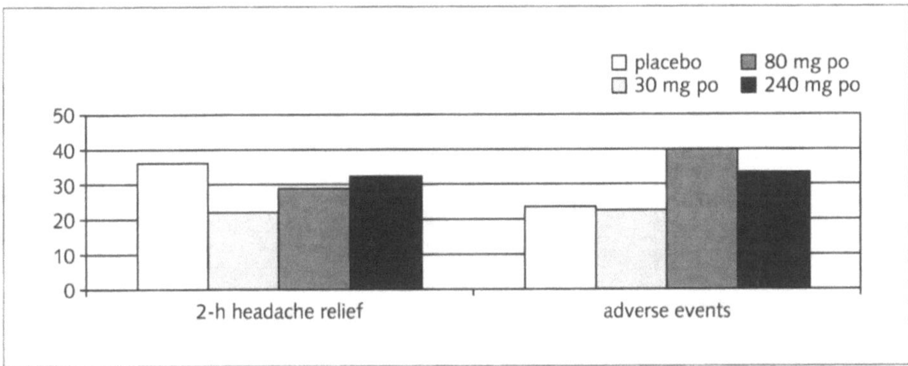

Figure 6
Response to LY-303870, 30, 80, and 240 mg orally, in percentage responders, compared to placebo, in the abortive treatment of moderate or severe migraine headache. Data obtained from [21].

occurrence of adverse events are presented in Figure 6. The results on headache relief revealed no benefit.

L-758298

L-758298 is a non-peptide, phosphate pro-drug of the brain-penetrant neurokinin-1-receptor antagonist, L-754030. Given intravenously in man, it has been shown to inhibit forearm vasodilation induced by substance P, without significant effects on heart rate and blood pressure [23]. It has been studied clinically in 72 patients, who either received 20, 40, or 60 mg of the compound intravenously or placebo [24]. The results for the 60-mg dose with regard to 2-h headache relief are shown in Figure 7; no information on the occurrence of adverse events is available. The results on headache relief revealed no benefit.

GR-205171

GR-205171 is a non-peptide neurokinin-1-receptor antagonist, which also penetrates the brain. It has been shown to inhibit plasma-protein extravasation in the dura mater and extracranial tissues of the rat, following electrical stimulation of the trigeminal ganglion [25]. It has been studied clinically in 63 patients, who received either 25 mg of the compound intravenously (n = 31) or placebo (n = 32) [26]. The

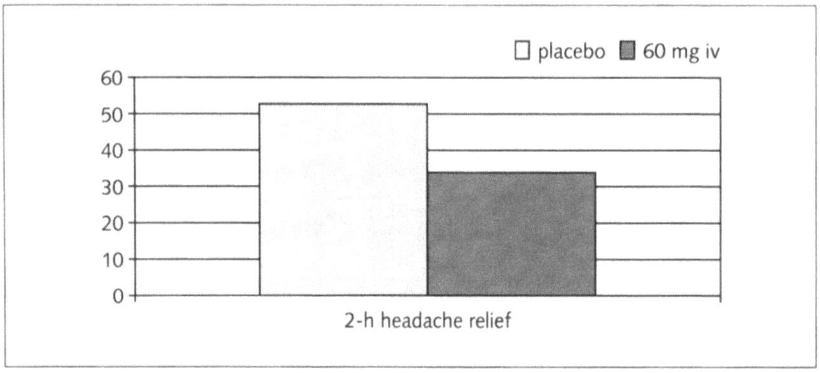

Figure 7
Response to L-758298, 60 mg intravenously, in percentage responders, compared to placebo, in the abortive treatment of moderate or severe migraine headache. Data obtained from [22].

information available from the study is that the compound was not effective in the abortive treatment of migraine.

FK-888

FK-888 is a peptide neurokinin-1-receptor antagonist of low molecular weight, which has been shown to lack carotid vasoconstrictor activity in the dog. It has been studied clinically, administered by inhalation in doses of 20, 40, and 80 mg. The results on headache relief revealed no benefit in comparison to placebo.

Conclusion

Of the 11 selective inhibitors of neurogenic inflammation studied clinically, only one, LY-334370, was found to be effective in the abortive treatment of migraine. However, this compound was only effective in relatively high doses, generating plasma concentrations at which interaction with serotonin-1B receptors cannot be excluded. An additional problem is that the focus of the compounds was on substance P and the plasma-protein extravasation mediated by this neuropeptide, although the prejunctionally acting compounds also inhibit the release of calcitonin gene-related peptide. Once substance P binds to and activates its receptor, the neurokinin-1 receptor, the complex is taken up into the target cell and is unavailable to antagonists present extracellularly. This is different with calcitonin gene-related peptide, which mediates the vasodilation component of neurogenic inflammation through a specific receptor present on the vascular smooth muscle. Studies with a selective antagonist of the receptor for calcitonin gene-related peptide, BIBN4096BS [27], are ongoing and the results are eagerly awaited.

There may still be a potential for the compounds in the preventive treatment of migraine, which certainly seems a more rational approach from a pathophysiological perspective. Also, the compounds have been generally well tolerated, with few if any side effects, which is important when they have to be taken daily for a longer period of time, as is the case in migraine prevention. So far, one of the selective, postjunctionally acting neurokinin-1-receptor antagonists, LY-303870 (lanepitant), has been studied preventively but with negative results [28]. It may still be worthwhile to subject other compounds to this clinical test as well, especially those acting prejunctionally and addressing the involvement of both substance P and calcitonin gene-related peptide. However, admittedly, with the above negative results in abortive treatment, it is unlikely that this potential will ever be further explored.

Acknowledgement
Richard J. Hargreaves, Ph.D., critically reviewed the manuscript and provided helpful comments.

References

1 Chapman LF, Ramos AO, Goodell H et al (1960) A humoral agent implicated in vascular headache of the migraine type. *Arch Neurol* 3: 223–229

2 Graham JR, Wolff HG (1938) Mechanism of migraine headache and action of ergotamine tartrate. *Arch Neurol Psychiat* 39: 737–763

3 Spierings ELH (2000) The (suma)triptan history revisited (letter). *Headache* 40: 766–767

4 Buzzie MG, Moskowitz MA (1990) The antimigraine drug, sumatriptan (GR43175), selectively blocks neurogenic plasma extravasation from blood vessels in dura mater. *Br J Pharmacol* 99: 202–206

5 Earl NL, McDonald SA, Lowy MT (1999) Efficacy and tolerability of the neurogenic inflammation inhibitor, 4991W93, in the acute treatment of migraine (abstract). *Cephalalgia* 19: 357

6 Mancilla BG, Cutler N, Leibowitz MT et al (2001) Safety and efficacy of PNU-142633, a selective 5-HT_{1D} agonist, in patients with acute migraine. *Cephalalgia; in press*

7 Johnson KW, Schaus JM, Durkin MM et al (1997) 5-HT_{1F} receptor agonists inhibit neurogenic dural inflammation in guinea pigs. *Neuroreport* 8: 2237–2240

8 Shepheard S, Edvinsson L, Cumberbatch M et al (1999) Possible antimigraine mechanisms of action of the $5HT_{1F}$ receptor agonist LY334370. *Cephalalgia* 19: 851–858

9 Goldstein DJ, Roon KI, Offen WW et al (1999) Migraine treatment with selective 5-HT_{1F} receptor agonist (SSOFRA) LY334370 (abstract). *Cephalalgia* 19: 318

10 Lee WS, Moskowitz MA (1993) Conformationally restricted sumatriptan analogues, CP-122,288 and CP-122,638, exhibit enhanced potency against neurogenic inflammation in dura mater. *Brain Res* 626: 303–305

11 Gupta P, Brown D, Butler P et al (1995) The in vivo pharmacological profile of a 5-HT_1 receptor agonist, CP-122,288, a selective inhibitor of neurogenic inflammation. *Br J Pharmacol* 116: 2385–2390

12 Shepheard SL, Williamson DJ, Beer MS et al (1997) Differential effects of 5-$HT_{1B/1D}$ receptor agonists on neurogenic dural plasma extravasation and vasodilation in anaesthetized rats. *Neuropharmacology* 36: 525–533

13 Roon KI, Olesen J, Diener HC et al (2000) No acute antimigraine efficacy of CP-122288, a highly potent inhibitor of neurogenic inflammation: results of two randomized, double-blind, placebo-controlled clinical trials. *Ann Neurol* 47: 238–241

14 Clozel M, Breu V, Gray GA (1994) Pharmacological characterization of bosetan, a new potent orally active non-peptide endothelin receptor antagonist. *J Pharmacol Exp Ther* 270: 228–235

15 Brändli P, Löffler BM, Breu V et al (1996) Role of endothelin in mediating neurogenic plasma extravasation in rat dura mater. *Pain* 64: 315–322

16 May A, Gijsman HJ, Wallnöfer et al (1996) Endothelin antagonist bosetan blocks neurogenic inflammation, but is not effective in aborting migraine attacks. *Pain* 67: 375–378

17 Smith MI, Read SJ, Chan WN et al (2000) Repetitive cortical spreading depression in a gyrencephalic feline brain: inhibition by the novel benzoylamino-benzopyran SB-220453. *Cephalalgia* 20: 546–553

18 Maassen van den Brink A, Van den Broek RWM, De Vries R et al (2000) The potential anti-migraine compound SB-220453 does not contract human isolated blood vessels or myocardium; a comparison with sumatriptan. *Cephalalgia* 20: 538–545

19 Moussaoui SM, Garret C (1997) Are substance P antagonists potential treatment for migraine and related headaches, pain, and emesis? In: J Olesen, P Tfelt-Hansen (eds): *Headache treatment: trial methodology and new drugs*. Lippincott-Raven Publishers, Philadelphia, 303–311

20 Lee WS, Moussaoui S, Moskowitz MA (1994) Blockade of oral or parenteral RPR 100893 (a non-peptide NK_1 receptor antagonist) of neurogenic plasma protein extravasation within guinea-pig dura mater and conjunctiva. *Br J Pharmacol* 112: 920–924

21 Phebus LA, Johnson KW, Stengel PW et al (1997) The non-peptide NK-1 receptor antagonist LY303870 inhibits neurogenic dural inflammation in guinea pigs. *Life Sciences* 60: 1553–1561

22 Goldstein DJ, Wang O, Saper JR et al (1997) Ineffectiveness of neurokinin-1 antagonists in acute migraine: a crossover study. *Cephalalgia* 17: 785–790

23 Newby DE, Sciberras DG, Ferro CJ et al (1999) Substance P-induced vasodilation is

mediated by the neurokinin type 1 receptor but does not contribute to basal vascular tone in man. *Br J Clin Pharmacol* 48: 336–344
24 Norman B, Panebianco D, Block GA (1998) A placebo-controlled, in-clinic study to explore the preliminary safety and efficacy of intravenous L-758,298 (a prodrug of the NK_1 receptor antagonist L-754,030) in the acute treatment of migraine (abstract). *Cephalalgia* 18: 407
25 Polley JS, Gaskin PJ, Perren MJ et al (1997) The activity of GR205171, a potent non-peptide tachykinin NK_1 receptor antagonist, in the trigeminovascular system. *Regulatory Peptides* 68: 23–29
26 Connor HE, Bertin L, Gillies S et al (1998) Clinical evaluation of a novel, potent, CNS penetrating NK_1 receptor antagonist in the acute treatment of migraine (abstract). *Cephalalgia* 18: 392
27 Doods H, Hallermayer G, Wu D et al (2000) Pharmacological profile of BIBN4096BS, the first selective small molecule CGRP antagonist. *Br J Pharmacol* 129: 420–423
28 Goldstein DJ, Offen WW, Klein EG, Phebus LA (1999) Lanepitant, an NK-1 antagonist, in migraine prophylaxis (abstract). *Cephalalgia* 19: 377

Inhibition of neurogenic inflammation in preventive migraine treatment

Margarita Sánchez del Río

Headache Unit, Department of Neurology, Fundación Hospital Alcorcón, Juan Carlos I University, Madrid, Spain

Introduction

Preventive pharmacological treatment of migraine aims at reducing the frequency, intensity and duration of the attacks by intervening, at the earliest stage possible, in migraine pathogenesis. The medications effective in migraine prevention conceivably act at multiple sites to raise the threshold for spreading cortical depression, to inhibit the mechanisms of vasodilation or neurogenic inflammation, or to enhance central anti-nociception. They most likely work through more than one mode of action and this review will focus on the evidence for an action in inhibiting neurogenic inflammation.

In the genesis of neurogenic inflammation two main structures are involved, blood vessels and nociceptive perivascular Aδ- and C-fibers of the trigeminal nerve. Chemical, electrical, or mechanical stimulation of the nerve fibers produces a sterile inflammation through the release of calcitonin gene-related peptide, neurokinin A and substance P [1]. The released neuropeptides activate endothelial cells, mast cells and blood platelets, causing vasodilation as well as an increase in extracellular amines, metabolites of arachidonic acid, peptides and ions [2, 3]. Metabolites of arachidonic acid, especially prostaglandins, play an important role in inflammation and the sensitization and pain associated with it. Enhanced release of prostaglandins has been documented after (antidromic) stimulation of sensory nerve fibers in the dura mater [1] and during migraine headache [4, 5]. Prostaglandin E_2, but not substance P, neurokinin A, or calcitonin gene-related peptide, causes sensitization of visceral sensory nerve fibers [6, 7]. Norepinephrine and prostaglandins released from sympathetic nerve fibers significantly contribute to plasma-protein extravasation and edema formation occurring with inflammation [8]. Therefore, medications that block cyclo-oxygenase activity and, thereby, prostaglandin synthesis, or that antagonize the effects of the sympathomimetic amines decrease inflammation and pain sensitization [9–12].

Among the chemicals involved in migraine pathogenesis, there is strong evidence for serotonin as an endogenous pain modulator and vasoactive substance [13]. The

Migraine: A Neuroinflammatory Disease?, edited by Egilius L.H. Spierings and Margarita Sánchez del Río

fact that preventive medications act at serotonin-1 and -2 receptors further supports the role of serotonin in the pathogenesis of migraine [14, 15]. Fourteen different serotonin receptors have been identified but only a few (1A, 2B and 7) have been implicated in the mode of action of preventive antimigraine medications [16].

The relevance of the serotonin-2B receptor in migraine pathogenesis is based on the following observations. First, the serotonin-2B receptor, located on endothelial cells of cerebral blood vessels, is coupled to nitric oxide synthase, causing the local formation of nitric oxide [14, 17, 18]. Nitric oxide activates and sensitizes perivascular trigeminal nerve fibers by increasing intracellular cyclic GMP [19] and induces neurotransmitter release, resulting in vasodilation, plasma-protein extravasation, hyperalgesia and pain [18, 20, 21]. Nitric oxide also contributes to the initiation and maintenance of hyperalgesia produced by prostaglandin E_2 [19]. Second, the serotonin-2B receptor is positively coupled to the release of arachidonic acid *via* activation of phospholipase A2 [22]. Third, activation of the serotonin-2B receptor negatively regulates the G protein-coupled serotonin-1B/D-receptor system through metabolites of arachidonic acid dependent on cyclo-oxygenase [22]. Furthermore, indomethacin blocks the inhibitory effect of activation of the serotonin-2B receptor on the serotonin-1B/D receptor. The selective serotonin-2B-receptor antagonists, LY-266097, LY-202146 and LY-272015, administered orally dose-dependently decrease plasma-protein extravasation in the dura mater of the guinea pig induced by m-chlorophenylpiperazine [13, 23, 24]. This observation suggests that stimulation of the serotonin-2B receptor results in the production of nitric oxide and release of neuropeptides, causing vasodilation and inflammation, the two key elements in the pathogenesis of the migraine headache [25].

The second major serotonin receptor involved in the mode of action of preventive antimigraine medications is the serotonin-7 receptor. It is the most recently identified member of the family of G protein-coupled serotonin receptors. Messenger RNA encoding for the serotonin-7 receptor is present in human dura mater, predominantly in the vascular smooth muscle and endothelial cells [14, 26]. The receptor has been shown to mediate cerebral vasodilation through an endothelium-independent mechanism [27] and has been implicated in the vasodilation of neurogenic inflammation [28]. It is important that there is a high correlation between the effective clinical doses of several preventive antimigraine medications (amitriptyline: pKi 7.0; methysergide: pKi 7.1) and their affinity as antagonists of this receptor [27]. The above findings have led to the notion that the serotonin-7 receptor plays a role in the regulation of blood vessel tone and that preventive antimigraine medications with affinity for the serotonin-7 receptor counteract vasodilation induced by serotonin.

The serotonin-1A receptor has also been indirectly implicated in the pathogenesis of migraine, although fewer data support a relevant role. This receptor is located on neurons, glia and ependymal cells and inhibits the synthesis, metabolism and release of serotonin [29, 30]. Although serotonin-1A receptors mediate serotonin-induced contractions of the canine basilar artery, these receptors are involved neither in the vasoconstrictor actions of abortive antimigraine medications nor in their inhibitory effects on neurogenic inflammation [31]. Serotonin-1A receptors located in the nucleus of the solitary tract and area postrema have been implicated in the nausea and vomiting occurring during migraine headache. More specifically, serotonin-1A receptors are involved in the regulation of body temperature, mood changes and anxiety, frequent co-morbidities seen with migraine [32–34] .

In animals, multiple compounds block neurogenic inflammation, some of which have been shown to be effective in migraine prevention (Tab. 1). The precise mode of action of each compound is not well characterized but all in different degrees inhibit venous vasodilation or release of vasoactive and inflammatory peptides. Most of the compounds are not devoid of central effects, but whether this is the basis for their efficacy in migraine prevention requires further studies.

Beta-blockers

Beta-blockers are medications of first choice in the preventive treatment of migraine; however, their mode of action is one of the least well understood [37, 65] . Of the beta-blockers shown to be effective in migraine prevention, propranolol is the only one in this class studied and shown to be ineffective in inhibiting neurogenic inflammation after acute or chronic administration [35].

Beta-blockers antagonize serotonin-1A and -2B receptors [14, 15, 66]. However, no correlation exists between receptor affinity and potency in terms of migraine prevention. For example, propranolol and timolol display moderate affinity for the serotonin-1A receptor, whereas pindolol, while ineffective in migraine prevention, displays greater affinity (pKi 4.5) [67–69]. A recent study, using positron emission tomography with α-[11c]-methyl-l-tryptophan in migraine patients during preventive treatment with propranolol, revealed increased serotonin synthesis throughout the brain, probably through an action on the serotonin-1A receptor [70]. Some beta-blockers are not exempt from central action. Stimulation of serotonin-1A receptors on serotonergic neurons in the raphe nuclei inhibits intrinsic cell firing. The inhibitory effect of serotonin-1A-receptor agonists can be blocked by propranolol. Beta-blockers also exert an effect on central catecholaminergic systems [71]. Once again, there is no correlation between a central locus of action and efficacy in migraine prevention (Tab. 2).

Part of their effectiveness is probably attributable to inhibition of norepinephrine release by blocking prejunctional beta-adrenergic receptors on peripheral blood vessels [72]. Some medications, such as propanolol, also have membrane-stabilizing activity (decrease of intracellular calcium influx, phospholipase A2 activity and synthesis of platelet-activating factor), independent of the beta-adrenergic blocking action [73], probably rendering the system less "excitable". The fact that no single

Table 1 - Medications studied in migraine prevention and in experimental models of neurogenic inflammation

Medication	Effective in migraine prevention	Clinical trials[1]	Therapeutic gain[2]	Effective in inhibiting NI	NI studies
Beta-blockers				no	TGS [35]
a. Propanolol	+	[36, 37]	10–76%		
b. Nadolol	+	[38]			
c. Timolol	+	[36]			
d. Atenolol	+	[39]			
e. Metropolol	+	[40]			
Non-steroidal anti-inflammatory analgesics				yes	TGS [41] SSS [42, 43]
a. Acetylsalicylic acid	+/–	[44–46]	20–30%		
b. Naproxen	+	[47–49]	33%		
Antidepressants				yes	Knee inflammation [50]
a. Amitriptyline	+	[51]	40%		
b. Nortriptyline	+				
c. Doxepin	+				
d. Fluoxetine	+	[52, 53]	25%		
Calcium-channel blockers				yes	TGS [54], a [55]
a. Verapamil	+	[56, 57]	18–49%		
b. Flunarizine	+	[58, 59]			
Methysergide	+	[60]	14–30%	yes	TGS [35]
Cyproheptadine	+		34%		
Pizotifen	+	[61]			
Valproate	+	[63, 64]	39–43%	yes	TGS [62]

[1]double-blind randomized placebo-controlled clinical trials; [2]percentage of difference from placebo in headache frequency; TGS, trigeminal ganglion stimulation; SSS, superior sagittal sinus stimulation; NI, neurogenic inflammation; a, other inflammatory model

action on a specific system accounts for their antimigraine preventive activity suggests that either a combination of properties is necessary or that beta-blockers act *via* different mechanisms unknown to date.

Table 2 - Beta-blockers effective in migraine prophylaxis

Drug	CNS penetration	MSA	Cardio-selective	PAA	affinity for $5\text{-HT}_{2B/C}$ receptor	affinity for 5-HT_{1A} receptor
Propanolol	yes	yes	no	no	high	medium
Nadolol	–	no	no	no	–	–
Timolol	yes	no	no	no	–	medium
Atenolol	poorly	no	yes	no	low	
Metoprolol	yes	no	yes	no	–	–
Practolol	poorly	no	yes	yes	–	–

PAA, partial agonist activity; MSA, membrane-stabilizing activity

Non-steroidal anti-inflammatory analgesics

Non-steroidal anti-inflammatory analgesics used for migraine prevention include aspirin [46], ketoprofen [74], mefenamic acid [75] and naproxen [48]. Until recently, non-steroidal anti-inflammatory analgesics were thought to act predominantly through inhibiting cyclo-oxygenase or lipoxygenase activity, stabilizing proteins and reducing the level of inflammatory metabolites of arachidonic acid. Most currently available non-steroidal anti-inflammatory analgesics inhibit cyclo-oxygenase-1, constitutionally active and responsible for the physiological production of prostaglandins, and cyclooxygenase-2 induced in inflammation [76, 77]. The well-known side-effects of the non-steroidal anti-inflammatory medications (gastric ulcer secondary to decreased prostaglandin and thromboxane formation) arise from the inhibition of cyclooxygenase-1. Non-steroidal anti-inflammatory medications show different affinities for cyclooxygenase-1 and -2. Aspirin, indomethacin and ibuprofen have a higher affinity for cyclooxygenase-1 than for -2, while diclophenac and naproxen seem to inhibit both isoforms with equal potency. No studies have yet looked at the selective cyclooxygenase-2 inhibitors, celecoxib and rofecoxib, in the preventive treatment of migraine.

There is increasing evidence that non-steroidal anti-inflammatory analgesics exert their therapeutic action also *via* other mechanisms. Aspirin is known to reduce platelet aggregation and, thus, indirectly affects the release of vasoactive substances, particularly serotonin. Aspirin and ibuprofen also interfere with NF-κB, a pro-inflammatory gene that promotes the release of prostaglandins [77, 78]. In addition, anti-inflammatory analgesics constrict cranial blood vessels through inhibiting the formation of vascular relaxants, such as prostaglandin E_2 [79, 80] or nitric oxide [81, 82]. Both chemicals have been shown to produce sensitization and elicit pain in man and animals.

In animals, aspirin and indomethacin reduce plasma-protein extravasation after stimulation of the trigeminal ganglion or administration of substance P [41], while both ketorolac and aspirin inhibit nociceptive processing in the brainstem following antidromic electrical stimulation [42]. In addition, other properties attributed to indomethacin are a decrease in intracranial pressure [83] and vasoconstriction of cerebral blood vessels, most likely through a mechanism related to nitric oxide [84]. In addition, indomethacin blocks the inhibitory effects of the activation of the serotonin-2B receptor over the serotonin-1B/1D receptor [22]. It is interesting that indomethacin has a similar structure to serotonin due to an indolic moiety, which may explain some indomethacin-induced behavioral changes in elderly patients [85].

Apart from the peripheral modes of action, non-steroidal anti-inflammatory analgesics are thought to act centrally as well. Probably this action involves inhibition of prostaglandin synthesis in brain neurons, prolongation of catecholamine and serotonin turnover, blockade of the release of serotonin in response to noxious stimuli, and opioid-mediated analgesia [76, 86, 87].

Methysergide, pizotifen and cyproheptadine

Methysergide, pizotifen and cyproheptadine are serotonergic medications long used for migraine prevention. The side-effects of the long-term use of methysergide pose a particular problem (for review, see reference 88). The efficacy of the medications is attributed to their serotonin-2A, -2B and -2C receptor antagonism. Although initially attributed to the serotonin-2A receptor antagonism, it is more likely to be serotonin-2A/C receptor antagonism based on the fact that selective and potent serotonin-2A antagonists are not effective in migraine prevention [89]. The effective doses of the medications correlate well with their affinity for the serotonin-2B and -2C receptors (methysergide: pKi 5.7; pizotifen: pKi 3.6; cyproheptadine: pKi 2.2) [88, 90].

Methysergide is useful for migraine prevention when chronically administered but not for the acute attack. Similarly, in animals, pre-treatment with chronic administration of methysergide, but not a single dose, blocked neurogenic inflammation in the dura mater, induced either by trigeminal stimulation or intravenous capsaicin [35]. This suggests that methysergide presynaptically inhibits the release of peptides from the perivascular sensory nerve fibers.

Methysergide, cyproheptadine and pizotifen are also serotonin-1A receptor agonists (pKi 59–200) [67], with a lower affinity than that shown for the serotonin-2A receptor. In animals, methysergide has been shown to contract cerebral blood vessels and decrease carotid blood flow through stimulation of serotonin-1B/D receptors [91, 92]. Additional antagonistic effects at histamine-1, cholinergic, α_1-adrenergic, α_2-adrenergic and dopamine receptors have also been reported [93].

Antidepressants

The tricyclic antidepressant, amitriptyline, has proven efficacy in migraine prevention, independent of its antidepressant effect. Its mode of action probably involves a complex interplay between several neurotransmitter systems and neuroreceptors. Selective serotonin re-uptake inhibitors are also frequently used for migraine prevention. Among them, fluoxetine has shown promising results in clinical trials [52, 53]. Tricyclic antidepressants are known to block the uptake of different endogenous amines (norepinephrine, serotonin and dopamine) and antagonize multiple neurotransmitter receptors (histamine, acetylcholine and glutamate). However, the specific mode of action in migraine prevention remains unknown. Either serotonin-2 or -7 receptor antagonism or ability to block serotonin re-uptake may conceivably account for the preventive antimigraine efficacy [67]. *In vivo* experiments have shown that chronic administration of fluoxetine, paroxetine and clomipramine desensitize serotonin-2B and -2C receptors. Another possible mechanism for the antinociceptive action seen in inflammatory hyperalgesic models is through antagonism of the receptor for n-methyl-d-aspartate (NMDA) [94].

Chlorpromazine and amitriptyline were reported to inhibit the sustained contraction of the basilar artery and the aggregation of platelets in a dose-dependent manner, most probably due to their interaction with calcium [95]. Other tricyclic medications with analgesic properties (desipramine, cyproheptadine and carbamazepine) interact with the function of glutamate receptors to reduce intracellular calcium elevations induced by N-methyl-D-aspartate in neuronal cultures at concentrations that are clinically relevant [96]. In an inflammatory model in synovial tissue culture, amitriptyline and fluoxetine inhibited the production of nitric oxide and prostaglandin E_2, two major mediators of pain and inflammation [50].

Valproate

The efficacy of valproate in migraine prevention has been demonstrated in several clinical trials [97]. The mode of action of valproate is complex and involves multiple systems:

A. Increase in the inhibitory neurotransmitter, gamma-aminobutyric acid (GABA) through:
   1. Stimulation of the synthetic enzyme, glutamic acid decarboxylase, and
   2. Inhibition of the degrading enzymes, gamma-aminobutyric acid aminotransferase and succinate semialdehyde dehydrogenase [98, 99].

B. Decrease in the brain levels of excitatory amino acids and excitability mediated by N-methyl-D-aspartate (NMDA).

C. Increase in the brain levels of glycine, taurine, 5-hydroxyindoleacetic acid, dopamine and homovanillic acid (for review, see [100]).

D. Membrane stabilization through modulation of calcium, potassium and sodium channels [101, 102].

Given the multiplicity of actions of valproate, its clinical efficacy may be based on central as well as peripheral mechanisms. Valproate has been shown to decrease plasma-protein extravasation following stimulation of the trigeminal ganglion or administration of substance P [103]. This effect is mediated through peripheral GABA A receptors, as bicuculline, a $GABA_A$-receptor antagonist, completely reversed it and muscimol, a $GABA_A$-receptor agonist, mimicked it. The fact that valproate attenuated neurogenic inflammation in animals depleted of C-fibers and sphenopalatine ganglionectomy completely blocks its effects suggests that $GABA_A$ receptors are expressed by parasympathetic nerve fibers innervating the dura mater [103]. Valproate also attenuates c-fos expression, a marker of neuronal activation, in the trigeminal nucleus caudalis after noxious stimulation produced by intracisternal administration of capsaicin [104]. Other inhibitory actions of valproate in the cerebral cortex or other central sites cannot be excluded.

Calcium-channel blockers

Calcium-channel blockers are extensively used in the preventive treatment of migraine. Flunarizine, acting on voltage-operated channels, and verapamil, acting on receptor-operated channels, are two medications with overlapping effects at the L-type calcium channel. A possible mode of action is the inhibition of calcium influx into the dural vascular smooth muscle and, consequently, the blockade of muscle contraction and vasoconstriction, irrespective of the constricting compound. Here, once more there is a lack of correlation between the potency of action of a medication and the clinical efficacy. Nimodipine, more potent than flunarizine at preventing calcium-induced contraction of human cerebral and temporal arteries, is less effective in migraine prevention. In addition, calcium-channel blockers reduce the inflammatory response by interfering with the expression of adhesion molecules, such as VCAM-1 and ICAM-1, in endothelial cells, reducing the migration of leukocytes and their adhesion to post-capillary venules [55]. Flunarizine has been shown to significantly inhibit neurogenic inflammation caused by electrical stimulation of the trigeminal ganglion, either by presynaptic inhibition of neuropeptide release or inhibition of other inflammatory responses [54].

Other suggested modes of action are blockade of platelet serotonin release and aggregation, as well as antagonism of α_2- and β_2-adrenergic, dopaminergic 2, histamine and serotonin-2 brain receptors [105, 106]. Interaction of flunarizine and verapamil with the opioid system and brain type IIA sodium channels has also been reported [107].

Table 3 - Modes of action at various prophylactic treatments of migraine

	β-blockers	Methys-ergide	Pizotifen	Cypro-heptadine	Valproate	Calcium-channel blockers	Amitrip-tyline	NSAIDs
A. On arterial tonus								
1. Reduce liability of vascular tonus	x	x				x		x
a.1 Serotonin receptors ($5\text{-}HT_{2B}$)/NO	x	x						x
a.2 Serotonin receptors ($5\text{-}HT_7$)		x					x	
B. On pain threshold								
1. Prevent disturbance in permeability								
a.1 Histamine release		x	x	x				
a.2 Histamine receptors			x	x		x	x	
b.1 Serotonin synthesis/release (5-HT-1A)	x	x				x	x	x
b.2 Serotonin receptors (5HT-2B)	x	x	x	x			x	x
c. Platelet aggregation		x			x	x		x
d. CGRP secretion					x			
2. Prevent plasmakinin formation			x					x
3. Prevent potentiation of prostaglandin-induced pain by serotonin		x						
4. $GABA_A$ receptors					x			
5. Decreasing glucose release					x			
6. Stabilization of membrane	x				x	x		
7. Opioid activity						x		x

Conclusion

The use of many of the preventive antimigraine medications is based on empirical reasons. Their effectiveness in decreasing the frequency and intensity of migraine attacks has justified their continuous use. This has also precluded the investigation of the subjacent mechanism, for practical reasons. So far, we know that the preventive medications work through more than one mechanism (Tab. 3). Inhibition of venous vasodilation or release of vasoactive and inflammatory peptides is one of the modes of action. Action of preventive medications on serotonin receptors (most medications), nitric oxide (indomethacin, methysergide, beta-blockers), cyclo-oxygenase (non-steroidal anti-inflammatory analgesics), gamma-aminobutyric acid (GABA) (valproate) and calcium channels (flunarizine, verapamil), among others, have revealed a relevant role of other neurotransmitters in the modulation of the transmission of pain originating from the head. As these medications also exert central nervous system effects, it is difficult to justify their action solely on a peripheral mechanism. Most likely a combination of central and peripheral mechanisms is involved, as demonstrated by the fact that the more "dirty" the medication is the more effective it is in migraine prevention.

References

1 Ebersberger A, Averbeck B, Messlinger K, Reeh PW (1999) Release of substance P, calcitonin gene-related peptide and prostaglandin E2 from rat dura mater encephali following electrical and chemical stimulation *in vitro*. *Neuroscience* 89: 901–907

2 Moskowitz MA (1993) Neurogenic inflammation in the pathophysiology and treatment of migraine. *Neurology* 43: S16–S20

3 Moskowitz MA (1992) Neurogenic *versus* vascular mechanisms of sumatriptan and ergot alkaloids in migraine. *Trends Pharmacol Sci* 13: 307–311

4 Tuca JO, Planas JM, Parellada PP (1989) Increase in PG_{E2} and TX_{A2} in the saliva of common migraine patients. Action of calcium channel blockers. *Headache* 29: 498–501

5 Puig-Parellada P, Planas JM, Gimenez J et al (1991) Plasma and saliva levels of PG_{I2} and TX_{A2} in the headache-free period of classical migraine patients. The effects of nicardipine. *Headache* 31: 156–158

6 Chen X, Tanner K, Levine JD (1999) Mechanical sensitization of cutaneous c-fiber nociceptors by prostaglandin E2 in the rat. *Neurosci Lett* 267: 105–108

7 Minami T, Okuda-Ashitaka E, Hori Y et al (1999) Involvement of primary afferent c-fibres in touch-evoked pain (allodynia) induced by prostaglandin E2. *Eur J Neurosci* 11: 1849–1856

8 Coggeshall RE, Carlton SM (1999) Evidence for an inflammation-induced change in the local glutamatergic regulation of postganglionic sympathetic efferents. *Pain* 83: 163–168

9 Ferreira SH (1993) The role of interleukins and nitric oxide in the mediation of inflammatory pain and its control by peripheral analgesics. *Drugs* 46 (Suppl) 1: 1–9

10 Nakamura M, Ferreira SH (1987) A peripheral sympathetic component in inflammatory hyperalgesia. *Eur J Pharmacol* 135: 145–153

11 Kinnman E, Levine JD (1995) Involvement of the sympathetic postganglionic neuron in capsaicin-induced secondary hyperalgesia in the rat. *Neuroscience* 65: 283–291

12 Khasar SG, Miao FJ, Levine JD (1995) Inflammation modulates the contribution of receptor-subtypes to bradykinin-induced hyperalgesia in the rat. *Neuroscience* 69: 685–690

13 Johnson KW, Phebus LA, Cohen ML (1998) Serotonin in migraine: theories, animal models and emerging therapies. *Prog Drug Res* 51: 219–244

14 Schmuck K, Ullmer C, Kalkman HO et al (1996) Activation of meningeal 5-HT_{2B} receptors: an early step in the generation of migraine headache? *Eur J Neurosci* 8: 959–967

15 Peroutka SJ (1990) Developments in 5-hydroxytryptamine receptor pharmacology in migraine. *Neurol Clin* 8: 829–839

16 Hoyer D, Clarke DE, Fozard JR et al (1994) International Union of Pharmacology classification of receptors for 5-hydroxytryptamine (serotonin). *Pharmacol Rev* 46: 157–203

17 Martin GR, Bolofo ML, Giles H (1992) Inhibition of endothelium-dependent vasorelaxation by arginine analogues: a pharmacological analysis of agonist and tissue dependence. *Br J Pharmacol* 105: 643–652

18 Wei EP, Moskowitz MA, Boccalini P, Kontos HA (1992) Calcitonin gene-related peptide mediates nitroglycerin and sodium nitroprusside-induced vasodilation in feline cerebral arterioles. *Circ Res* 70: 1313–1319

19 Aley KO, McCarter G, Levine JD (1998) Nitric oxide signaling in pain and nociceptor sensitization in the rat. *J Neurosci* 18: 7008–7014

20 Yonehara N, Yoshimura M (1999) Effect of nitric oxide on substance P release from the peripheral endings of primary afferent neurons. *Neurosci Lett* 271: 199–201

21 Garry MG, Richardson JD, Hargreaves KM (1994) Sodium nitroprusside evokes the release of immunoreactive calcitonin gene-related peptide and substance P from dorsal horn slices *via* nitric oxide-dependent and nitric oxide-independent mechanisms. *J Neurosci* 14: 4329–4337

22 Tournois C, Mutel V, Manivet P et al (1998) Cross-talk between 5-hydroxytryptamine receptors in a serotonergic cell line. Involvement of arachidonic acid metabolism. *J Biol Chem* 273: 17498–17503

23 Bonhaus DW, Flippin LA, Greenhouse RJ et al (1999) RS-127445: a selective, high affinity, orally bioavailable 5-HT_{2B} receptor antagonist. *Br J Pharmacol* 127: 1075–1082

24 Nelson AD, Wainscott DB, Lucaites VL et al (1997) Selective $5HT_{2B}$ receptor antagonists block mCPP-induced dural extravasation (abstract). *Cephalalgia* 17: 342

25 Moskowitz MA, Macfarlane R (1993) Neurovascular and molecular mechanisms in migraine headaches. *Cerebrovasc Brain Metab Rev* 5: 159–177

26 Cohen Z, Bouchelet I, Olivier A et al (1999) Multiple microvascular and astroglial 5-

hydroxytryptamine receptor subtypes in human brain: molecular and pharmacologic characterization. *J Cereb Blood Flow Metab* 19: 908–917
27 Terron JA (1998) Involvement of the 5-HT_7 receptor in craniovascular vasodilatation: potential impact in migraine. *Proc West Pharmacol Soc* 41: 247–251
28 Terron JA, Falcon-Neri A (1999) Pharmacological evidence for the 5-HT_7 receptor mediating smooth muscle relaxation in canine cerebral arteries. *Br J Pharmacol* 127: 609–616
29 Azmitia EC, Gannon PJ, Kheck NM, Whitaker-Azmitia PM (1996) Cellular localization of the 5-HT_{1A} receptor in primate brain neurons and glial cells. *Neuropsychopharmacology* 14: 35–46
30 Peroutka SJ (1988) Antimigraine drug interactions with serotonin receptor subtypes in human brain. *Ann Neurol* 23: 500–504
31 Moskowitz MA, Waeber C (1997) Neuronal pathophysiology of migraine as a basis for acute treatment with 5-HT receptor ligands. In: H Baumgarten, M Gother (eds): *Handbook of experimental pharmacology*, volume 128. Springer Verlag, Berlin, 613–635
32 Cryan JF, Kelliher P, Kelly JP, Leonard BE (1999) Comparativc cffects of serotonergic agonists with varying efficacy at the 5-HT(1A) receptor on core body temperature: modification by the selective 5-HT(1A) receptor antagonist WAY 100635. *J Psychopharmacol* 13: 278–283
33 Ebenezer IS, Parrott RF, Vellucci SV (1999) Effects of the 5-HT_{1A} receptor agonist 8-OH-DPAT on operant food intake in food-deprived pigs. *Physiol Behav* 67: 213–217
34 Esteban S, Llado J, Sastre-Coll A, Garcia-Sevilla JA (1999) Activation and desensitization by cyclic antidepressant drugs of alpha 2- autoreceptors, alpha 2-heteroreceptors and 5-HT_{1A}-autoreceptors regulating monoamine synthesis in the rat brain *in vivo*. *Naunyn Schmiedebergs Arch Pharmacol* 360: 135–143
35 Markowitz S, Saito K, Moskowitz MA (1988) Neurogenically mediated plasma extravasation in dura mater: effect of ergot alkaloids. A possible mechanism of action in vascular headache. *Cephalalgia* 8: 83–91
36 Tfelt-Hansen P, Standnes B, Kangasniemi P et al (1984) Timolol vs propranolol vs placebo in common migraine prophylaxis: a double-blind multicenter trial. *Acta Neurol Scand* 69: 1–8
37 Carroll JD, Reidy M, Savundra PA et al (1990) Long-acting propranolol in the prophylaxis of migraine: a comparative study of two doses. *Cephalalgia* 10: 101–105
38 Ryan RE (1984) Comparative study of nadolol and propranolol in prophylactic treatment of migraine. *Am Heart J* 108: 1156–1159
39 Stensrud P, Sjaastad O (1980) Comparative trial of Tenormin (atenolol) and Inderal (propranolol) in migraine. *Headache* 20: 204–207
40 Olsson JE, Behring HC, Forssman B et al (1984) Metoprolol and propranolol in migraine prophylaxis: a double-blind multicenter study. *Acta Neurol Scand* 70: 160–168
41 Buzzi MG, Sakas DE, Moskowitz MA (1989) Indomethacin and acetylsalicylic acid block neurogenic plasma protein extravasation in rat dura mater. *Eur J Pharmacol* 165: 251–258

42 Kaube H, Hoskin KL, Goadsby PJ (1993) Intravenous acetylsalicylic acid inhibits central trigeminal neurons in the dorsal horn of the upper cervical spinal cord in the cat. *Headache* 33: 541–544

43 Ellrich J, Schepelmann K, Pawlak M, Messlinger K (1999) Acetylsalicylic acid inhibits meningeal nociception in rat. *Pain* 81: 7–14

44 Diener HC (1999) Efficacy and safety of intravenous acetylsalicylic acid lysinate compared to subcutaneous sumatriptan and parenteral placebo in the acute treatment of migraine. A double-blind, double-dummy, randomized, multicenter, parallel group study. *Cephalalgia* 19: 581–588

45 Lipton RB, Stewart WF, Ryan RE et al (1998) Efficacy and safety of acetaminophen, aspirin, and caffeine in alleviating migraine headache pain: three double-blind, randomized, placebo-controlled trials. *Arch Neurol* 55: 210–217

46 Buring JE, Peto R, Hennekens CH (1990) Low-dose aspirin for migraine prophylaxis. *J Am Med Assoc* 264: 1711–1713

47 Sances G, Martignoni E, Fioroni L et al (1990) Naproxen sodium in menstrual migraine prophylaxis: a double-blind placebo controlled study. *Headache* 30: 705–709

48 Welch KM, Ellis DJ, Keenan PA (1985) Successful migraine prophylaxis with naproxen sodium. *Neurology* 35: 1304–1310

49 Ziegler DK, Ellis DJ (1985) Naproxen in prophylaxis of migraine. *Arch Neurol* 42: 582–584

50 Yaron I, Shirazi I, Judovich R et al (1999) Fluoxetine and amitriptyline inhibit nitric oxide, prostaglandin E2, and hyaluronic acid production in human synovial cells and synovial tissue cultures. *Arthritis Rheum* 42: 2561–2568

51 Couch JR, Hassanein RS (1979) Amitriptyline in migraine prophylaxis. *Arch Neurol* 36: 695–699

52 Oguzhanoglu A, Sahiner T, Kurt T, Akalin O (1999) Use of amitriptyline and fluoxetine in prophylaxis of migraine and tension-type headaches. *Cephalalgia* 19: 531–532

53 Steiner TJ, Ahmed F, Findley LJ et al (1998) S-fluoxetine in the prophylaxis of migraine: a phase II double-blind randomized placebo-controlled study. *Cephalalgia* 18: 283–286

54 Hashimoto M, Yamamoto Y, Takagi H (1997) Effects of KB-2796 on plasma extravasation following antidromic trigeminal stimulation in the rat. *Res Commun Mol Pathol Pharmacol* 97: 79–94

55 Martinez LL, Aparecida DO, Fortes ZB (1999) Influence of verapamil and diclofenac on leukocyte migration in rats. *Hypertension* 34: 997–1001

56 Markley HG, Cheronis JC Piepho RW (1984) Verapamil in prophylactic therapy of migraine. *Neurology* 34: 973–976

57 Solomon GD, Steel JG, Spaccavento LJ (1983) Verapamil prophylaxis of migraine. A double-blind, placebo-controlled study. *J Am Med Assoc* 250: 2500–2502

58 Sørensen PS, Hansen K, Olesen J (1986) A placebo-controlled, double-blind, cross-over trial of flunarizine in common migraine. *Cephalalgia* 6: 7–14

59 Frenken CW, Nuijten ST (1984) Flunarizine, a new preventive approach to migraine. A double-blind comparison with placebo. *Clin Neurol Neurosurg* 86: 17–20

60 Ryan RE (1968) Double-blind crossover comparison of bc-105, methysergide and placebo in the prophylaxis of migraine headache. *Headache* 8: 118–126
61 Gillies D, Sills M, Forsythe I (1986) Pizotifen (Sanomigran) in childhood migraine. A double-blind controlled trial. *Eur Neurol* 25: 32–35
62 Lee WS, Limmroth V, Ayata C et al (1995) Peripheral GABA-A receptor-mediated effects of sodium valproate on dural plasma protein extravasation to substance P and trigeminal stimulation. *Br J Pharmacol* 116: 1661–1667
63 Klapper J (1997) Divalproex sodium in migraine prophylaxis: a dose-controlled study. *Cephalalgia* 17: 103–108
64 Hering R, Kuritzky A (1992) Sodium valproate in the prophylactic treatment of migraine: a double-blind study *versus* placebo. *Cephalalgia* 12: 81–84
65 Pradalier A, Serratrice G, Collard M et al (1989) Long-acting propranolol in migraine prophylaxis: results of a double-blind, placebo-controlled study. *Cephalalgia* 9: 247–253
66 Martin GR (1994) Vascular receptors for 5-hydroxytryptamine: distribution, function and classification. *Pharmacol Ther* 62: 283–324
67 Peroutka SJ (1990) The pharmacology of current anti-migraine drugs. *Headache* 30: 5–11
68 Gobert A, Millan MJ (1999) Modulation of dialysate levels of dopamine, noradrenaline, and serotonin (5-HT) in the frontal cortex of freely-moving rats by (–) pindolol alone and in association with 5-HT reuptake inhibitors: comparative roles of beta-adrenergic, 5-HT_{1A}, and 5-HT_{1B} receptors. Neuropsychopharmacology 21: 268–284
69 Newman-Tancredi A, Chaput C, Gavaudan S et al (1998) Agonist and antagonist actions of (–)pindolol at recombinant, human serotonin 1A (5-HT_{1A}) receptors. *Neuropsychopharmacology* 18: 395–398
70 Chugani DC, Niimura K, Chaturvedi S (1999) Increased brain serotonin synthesis in migraine. *Neurology* 53: 1473–1479
71 Schoenen J, Maertens DN, Timsit-Berthier M, Timsit M (1986) Contingent negative variation and efficacy of beta-blocking agents in migraine. *Cephalalgia* 6: 229–233
72 Saxena PR, Heiligers JP, Villalon CM, Ferrari MD (1992) Effects of tertatolol, a beta-adrenoceptor antagonist with agonist affinity at 5-HT_{1A} receptors, in an animal model of migraine: comparison with propranolol and pindolol. *Eur J Pharmacol* 220: 79–86
73 Anderson R, Ramafi G, Theron AJ (1996) Membrane stabilizing, anti-oxidative interactions of propranolol and dexpropranolol with neutrophils. *Biochem Pharmacol* 52: 341–349
74 Stensrud P, Sjaastad O (1974) Clinical trial of a new anti-bradykinin, anti-inflammatory drug, ketoprofen in migraine prophylaxis. *Headache* 14: 96–100
75 Johnson RH, Hornabrook RW, Lambie DG (1986) Comparison of mefenamic acid and propranolol with placebo in migraine prophylaxis. *Acta Neurol Scand* 73: 490–492
76 Cashman JN (1996) The mechanisms of action of NSAIDs in analgesia. *Drugs* 52 (Suppl 5): 13–23

77 Kurumbail RG, Stevens AM, Gierse JK et al (1996) Structural basis for selective inhibition of cyclooxygenase-2 by anti-inflammatory agents. *Nature* 384: 644–648

78 Stuhlmeier KM, Li H, Kao JJ (1999) Ibuprofen: new explanation for an old phenomenon. *Biochem Pharmacol* 57: 313–320

79 Carroll JD, Reidy M, Savundra PA et al (1990) Long-acting propranolol in the prophylaxis of migraine: a comparative study of two doses. *Cephalalgia* 10: 101–105

80 Rouaud C, Delaforge M, Anger-Leroy M et al (1999) The cyclo-oxygenase-dependent regulation of rabbit vein contraction: evidence for a prostaglandin E2-mediated relaxation. *Br J Pharmacol* 126: 35–44

81 Rhodes KF, Buckingham JC, Kennard C (1999) The inhibition of nicotine-evoked relaxation of the guinea-pig isolated basilar artery by some analgesic drugs and progesterone. *Br J Pharmacol* 126: 1003–1008

82 Chen X, Levine JD (1999) NOS inhibitor antagonism of PG_{E2}-induced mechanical sensitization of cutaneous c-fiber nociceptors in the rat. *J Neurophysiol* 81: 963–966

83 Harrigan M R, Tuteja S, Neudeck B L (1997) Indomethacin in the management of elevated intracranial pressure: a review. *J Neurotrauma* 14: 637–650

84 Castellano A E, Micieli G, Bellantonio P et al. (1998) Indomethacin increases the effect of isosorbide dinitrate on cerebral hemodynamic in migraine patients: pathogenetic and therapeutic implications. *Cephalalgia* 18: 622–630

85 Mallet L,Kuyumjian J (1998) Indomethacin-induced behavioral changes in an elderly patient with dementia. *Ann Pharmacother* 32: 201–203

86 Bromm B, Rundshagen I, Scharein E (1991) Central analgesic effects of acetylsalicylic acid in healthy men. *Arzneimittelforschung* 41: 1123–1129

87 Attal N, Kayser V, Eschalier A et al (1988) Behavioural and electrophysiological evidence for an analgesic effect of a non-steroidal anti-inflammatory agent, sodium diclofenac. *Pain* 35: 341–348

88 Silberstein SD (1998) Methysergide. *Cephalalgia* 18: 421–435

89 Martucci N, Manna V, Agnoli A (1985) Antidepressant drugs and migraine. *Cephalalgia* 5 (Suppl 2): 225–228

90 Kalkman HO (1994) Is migraine prophylactic activity caused by 5-HT_{2B} or 5-HT_{2C} receptor blockade? *Life Sci* 54: 641–644

91 Saxena PR, Verdouw PD (1984) Effects of methysergide and 5-hydroxytryptamine on carotid blood flow distribution in pigs: further evidence for the presence of atypical 5-HT receptors. *Br J Pharmacol* 82: 817–826

92 Villalon CM, De Vries P, Rabelo G et al (1999) Canine external carotid vasoconstriction to methysergide, ergotamine and dihydroergotamine: role of 5-$HT_{1B/1D}$ receptors and alpha 2-adrenoceptors. *Br J Pharmacol* 126: 585–594

93 Mylecharane EJ (1991) 5-HT_2 receptor antagonists and migraine therapy. *J Neurol* 238 (Suppl 1): S45–S52

94 Eisenach JC, Gebhart GF (1995) Intrathecal amitriptyline acts as an n-methyl-d-aspartate receptor antagonist in the presence of inflammatory hyperalgesia in rats. *Anesthesiology* 83: 1046–1054

95 Sukenaga A, Tani E, Fukumori T, Maeda Y (1984) Responses of isolated canine basilar artery and human platelet to chlorpromazine and amitriptyline. Stroke 15: 295–300
96 Cai Z, McCaslin PP (1992) Amitriptyline, desipramine, cyproheptadine and carbamazepine, in concentrations used therapeutically, reduce kainate- and N-methyl-D-aspartate-induced intracellular Ca^{2+} levels in neuronal culture. *Eur J Pharmacol* 219: 53–57
97 Rothrock JF (1997) Clinical studies of valproate for migraine prophylaxis. *Cephalalgia* 17: 81–83
98 Loscher W, Bohme G, Schafer H, Kochen W (1981) Effect of metabolites of valproic acid on the metabolism of GABA in brain and brain nerve endings. *Neuropharmacology* 20: 1187–1192
99 Loscher W (1993) *In vivo* administration of valproate reduces the nerve terminal (synaptosomal) activity of GABA aminotransferase in discrete brain areas of rats. *Neurosci Lett* 160: 177–180
100 Cutrer FM, Limmroth V, Moskowitz MA (1997) Possible mechanisms of valproate in migraine prophylaxis. *Cephalalgia* 17: 93–100
101 Van den Berg RJ, Kok P, Voskuyl RA (1993) Valproate and sodium currents in cultured hippocampal neurons. *Exp Brain Res* 93: 279–287
102 Van Erp MG, Van Dongen AM, Van den Berg RJ (1990) Voltage-dependent action of valproate on potassium channels in frog node of Ranvier. *Eur J Pharmacol* 184: 151–161
103 Lee W S, Limmroth V, Ayata C et al (1995) Peripheral $GABA_A$ receptor-mediated effects of sodium valproate on dural plasma protein extravasation to substance P and trigeminal stimulation. *Br J Pharmacol* 116: 1661–1667
104 Cutrer FM, Limmroth V, Ayata G, Moskowitz MA (1995) Attenuation by valproate of c-fos immunoreactivity in trigeminal nucleus caudalis induced by intracisternal capsaicin. *Br J Pharmacol* 116: 3199–3204
105 Peroutka SJ (1997) Dopamine and migraine. *Neurology* 49: 650–656
106 Greenberg DA (1986) Calcium channel antagonists and the treatment of migraine. *Clin Neuropharmacol* 9: 311–328
107 Weizman R, Getslev V, Pankova IA et al (1999) Pharmacological interaction of the calcium channel blockers verapamil and flunarizine with the opioid system. *Brain Res* 818: 187–195

Is migraine a neuroinflammatory disease?

Margarita Sánchez del Río

Headache Unit, Department of Neurology, Fundación Hospital Alcorcón, Juan Carlos I University, Madrid, Spain

The answer to the question raised in the title is complex and subject to ongoing debate. Acute inflammation is a well-characterized process that develops in vascularized connective tissue as a protective response, closely intertwined with the process of repair and independent of the eliciting stimulus. Inflammation and repair may be potentially harmful, as is the case in rheumatoid arthritis or even Alzheimer's disease. Acute inflammation has three major components:

1. Alterations in vascular caliber that lead to increased blood flow.
2. Structural microvasculature changes resulting in increased permeability and exudation of protein-rich fluid into the interstitium, the hallmark of inflammation.
3. Emigration of leukocytes.

It is important to keep in mind that the inflammatory response observed in migraine lacks the presence of leukocytes and is without evidence of tissue damage.

In migraine, inflammatory peptides, that is, calcitonin gene-related peptide, substance P and neurokinin A, are released from perivascular nerve fibers to promote vasodilation, activation of endothelium resulting in increased transendothelial transport, activation of mast cells, and enhanced local aggregation and adhesion of blood platelets. All these effects lead to the extravasation of plasma proteins in the dura mater. In this inflammatory process, the nociceptive Aδ- and C-fibers in the trigeminal nerve play a key role as the source of the inflammatory peptides. As in other inflammatory disorders, inflammation in migraine is neither the origin nor the leading event but one of the mechanisms involved in its pathogenesis, explaining findings such as sensitization.

In the preceding chapters, the role of neurogenic inflammation in the pathogenesis of migraine has been approached from different aspects, showing controversial results. Controversy in medicine leads to new research, provides a challenge to investigators and, it is hoped, will result in therapeutic benefits for patients. In the following, I will briefly review the scientific evidence that argues in favor of neurogenic inflammation playing a role in migraine and what evidence argues against.

Migraine: A Neuroinflammatory Disease?, edited by Egilius L.H. Spierings and Margarita Sánchez del Río

Findings in favor of neurogenic inflammation

1. Neurogenic inflammation provides for a mechanism to explain the occurrence of increased intracranial sensitivity to mechanical stimulation and hyperalgesia, often experienced as acute worsening of the pain by coughing or sudden head movement. Sensitization of sensory nerve fibers to previously innocuous stimuli (for example, pulsations of blood vessels or changes in venous pressure) may underlie this phenomenon, as shown in experimental animals by Strassman et al. [1]. Chemical sensitization of C-fibers after topical application of inflammatory soup decreased the threshold to mechanical stimuli of neurons, previously responsive to high-level stimulation only or previously unresponsive to mechanical stimuli. A similar phenomenon has been demonstrated in migraineurs where cutaneous allodynia is observed inside and outside the painful area during migraine headache [2–4]. Even more, the recruitment of areas with allodynia followed a temporal pattern, suggesting progression of the sensitization process from peripheral to central structures [4].
2. A preliminary study using single-photon positron emission computerized tomography performed with serum albumin labeled with 99-technetium, injected several hours after the onset of a migraine headache, revealed increased activity in localized extraparenchymal regions ipsilateral to the pain, suggestive of dural inflammation, without leakage occurring in the brain tissue [5]. The findings parallel those observed in animals using similar techniques to demonstrate plasma-protein extravasation. The fact that extravasation is localized to the site of pain and in an area corresponding to the dura illustrates how activation of specific dural branches of the trigeminal nerve plays a key role in the development of neurogenic inflammation in migraine.

Findings against neurogenic inflammation

1. Neuroimaging techniques using conventional magnetic resonance imaging have repeatedly failed to demonstrate the occurrence of contrast extravasation in the dura mater during migraine headache. Although this may argue against the occurrence of inflammation, it must be kept in mind that the technique is very limited in detecting small changes in a thin structure like the dura mater.
2. Studies of the human retina during migraine and cluster headache have failed to demonstrate plasma-protein extravasation [6]. In animals, stimulation of the trigeminal ganglion results in plasma-protein extravasation in the retinal choroid, a highly vascularized structure. The absence of contrast extravasation in the retina during migraine or cluster headache could be due to the difficulty of visualizing edema in the choroid because of its dense vascularization. Anoth-

er reason could be that only specific branches of the trigeminal nerve, that is, the dural branches, are involved in headache genesis.

3. Compounds targeted at a specific receptor site involved in neurogenic inflammation, that is, the neurokinin-1 receptor, are ineffective in aborting migraine headache. However, why should a compound acting at one particular receptor site be effective when several neurotransmitters are simultaneously involved? This is further supported by the clinical experience that all effective antimigraine medications have more than one mode of action.

Most of the findings that argue against neurogenic inflammation do so in relation to the specific role that it plays in the pathogenesis of the migraine headache but not its occurrence *per se*. Those who defend the importance of neurogenic inflammation in the pathogenesis of migraine argue that it is not the origin of pain but rather that it is its presence that is fundamental to the genesis of the headache [7].

References

1 Strassman AM, Raymond SA, Burstein R (1996) Sensitization of meningeal sensory neurons and the origin of headaches. *Nature* 384: 560–564

2 Burstein R, Yamamura H, Malick A, Strassman AM (1998) Chemical stimulation of the intracranial dura induces enhanced responses to facial stimulation in brain stem trigeminal neurons. *J Neurophysiol* 79: 964–982

3 Burstein R, Yarnitsky D, Goor-Aryeh I et al (2000) An association between migraine and cutaneous allodynia. *Ann Neurol* 47: 614–624

4 Burstein R, Cutrer MF, Yarnitsky D (2000) The development of cutaneous allodynia during a migraine attack. Clinical evidence for the sequential recruitment of spinal and supraspinal nociceptive neurons in migraine. *Brain* 123: 1703–1709

5 Pappagalo M, Szabo Z, Esposito G et al (1999) Imaging neurogenic inflammation in patients with migraine headaches (abstract). *Neurology* 52: A274–A275

6 May A, Shepheard SL, Knorr M et al (1998) Retinal plasma extravasation in animals but not in humans: implications for the pathophysiology of migraine. *Brain* 121: 1231–1237

7 Sanchez del Rio M, Moskowitz MA (2000), The trigeminal system. In: J Olesen, P Tfelt-Hansen, KMA Welch (eds): *The headaches*. Lippincott Williams & Wilkins, Philadelphia, 141–149

Index

abortive antimigraine medication 123
absenteeism 25, 29
acetylcholine 74
adhesion molecule, endothelial 122
α-adrenergic-2 receptor 101
Aδ-fiber 91, 161
albumin, 99-technetium labeled 76
allodynia 4, 77, 162
allopregnanolone 74
γ-aminobutyric acid 73, 99
amitriptyline 151
amyl nitrite 6, 7
analgesic, non-steroidal anti-inflammatory 71, 149
antalarmin 119, 123
antidepressant, tricyclic 151
antidromic trigeminal ganglion stimulation 119
antidromic vasodilation 2
anxiety 23
arteriovenous anastomoses 17, 78, 133
ascending reticular activating system 12
associated migraine symptom 12
avitriptan 60
axon reflex flare 2
azatadine 119, 124

baclofen 74
benzoylamino-benzopyran compound 137
beta-blocker 147
BI-BN4096BS 141
bicuculline 74
blood oxygenation-level dependant signal 8, 9, 80
BMS-181885 60
bosetan 59, 137
brain arteriole, receptors for corticotrophin-releasing hormone 122
brain spectroscopy 10
brainstem 12, 13

calcitonin gene-related peptide 2, 4, 5, 18, 55, 71, 92, 96, 119, 133, 141, 161
calcium channelopathy 12
calcium-channel blocker 152
capsaicin 12, 68, 101, 123, 133
carbon dioxide 6
5-carboxamide tryptamine 17, 136
carotid blood flow 17
cat saphenous vein 17
catecholaminergic nerve fiber 115
cavernous sinus 76
cerebral blood flow 7, 8, 12
cerebral circulation 10, 11
cerebral cortex 11
cerebral vasoconstriction 6, 10, 11
cerebral vasodilation 6, 7, 10, 11
cerebrospinal fluid 1, 5, 6, 119
cervical ganglion stimulation 121
C-fiber 91, 121, 161
c-fos expression 69, 80
chiropractic manipulation 27
chlorophenylpiperazine 98

cholinergic nerve fiber 115
chronic paroxysmal hemicrania 75
cingulate cortex 12, 13
cluster headache 76, 121
cortical spreading depression 65, 79, 98
corticotrophin-releasing hormone 119, 122–124
CP-122288 59, 97, 135–137
cutaneous allodynia 77, 162
cyclic AMP 72
cyclo-oxygenase 1 100, 149
cyproheptadine 150

depression 23, 65, 79, 98
diencephalon 120
diffusion-weighted echo-planar magnetic resonance imaging 79
dihydroergotamine 14, 16, 17
disodium cromoglycate 120
dog saphenous vein 17
dorsal horn 122
dorsal root ganglion 122
dura mater 65, 115, 133
dural blood vessel, diameter 79
dural inflammation 162

electrical stimulation 69, 77
emergency room use 27
emotional stress 119
endothelial adhesion molecule 122
endothelin 59, 137
endothelin receptor 137
endothelin-receptor antagonist 137
enkephalin 5, 6
epidemiology of migraine 21
epilepsy 23
episodic ataxia type 2 80
ergot 14, 17, 18, 70
ergot, modes of action 14
ergotamine 1, 4, 14, 15, 133
exercise-induced migraine aura 9
external jugular vein 4, 5, 17, 18
extracranial arterial vasodilation 1, 10
extracranial artery 1, 3, 10, 15
extracranial tissue 11, 139
extracranial vasodilation 12
extravasation receptor 136

familial hemiplegic migraine 80
fiberoptic probe 122
FK-888 140
flunarizine 152
fluoxetine 151
flushing 122

GABA-A receptor 100
gamma-aminobutyric acid (GABA) 151
gluceptate, 99-technetium 120
glutamate 80
glutamic acid decarboxylase 74
GR-205171 78, 139
granulocyte infiltration 122

headache, daily 23
hemicrania, chronic paroxysmal 75
hemiplegic migraine, familial 80
histamine/serotonin-receptor antagoist 119, 124
histamine-3 receptor 101, 119
histamine-3-receptor agonist 124
hospitalization 27
hyperalgesia 162
hyperemia, reactive 8
hypothalamic-pituitary-adrenal axis 119
hypothalamus 115

ICAM-1 152
IL-2-receptor 122
IL-6 119
immobilization stress 119, 124
indomethacin 150
inflammation 2, 10–12, 14, 18, 56, 65, 76, 77, 91, 115, 119, 124, 133, 162
inflammatory peptide, source 161
inflammatory soup 76

interleukin (IL)-3, -4 and -6 115
intracranial arteries 1
intracranial histamine 115
intracranial sensitivity, increased 162
intravital microscopy 79
ischemia 80
itching 122

L-758298 139, 140
lanepitant 138, 141
laser beam excitation 122
leaner mouse 80
leukemic mast cell 123
leukocyte 122
limbic system 12
localization of mast cells, with neurons 115
LY-303870 (lanepitant) 138, 139, 141
LY-334370 97, 135, 136, 140

mania 23
mast cell, allergic reaction 115
mast cell hyperplasia 121
mast cell stabilizer 120
mastocytosis 121
meninge 115
mesencephalon 12–14
meta-chlorophenylpiperazine 98
n^{α}-methylhistamine 119
methysergide 150
migraine abortion, pharmacology 14
migraine attack 10, 11
migraine attack, pathogenesis 10
migraine aura 6, 8–12, 65, 79
migraine aura, pathogenesis 6
migraine aura without headache 12
migraine generator 12, 13
migraine headache 1, 10, 13–15, 18, 119
migraine headache, pathogenesis 1, 10, 11, 124
migraine mechanism 65
migraine pathogenesis, parallel concept 11
migraine pathogenesis, sequential concept 10
migraine prevention 141, 147
migraine process 11, 12
migraine trigger 11, 13
migraine with aura 1, 7–11
migraine without aura 8, 10–12
migrainous infarction 10
migrainous vasodilation 2
mononuclear leukocyte 122
muscimol 74

n-methyl-d-aspartate (NMDA) 151
nerve fiber 115
nerve growth factor 99
neurogenic inflammation 2, 10–12, 14, 18, 56, 65, 91, 119, 124, 133
neurogenic vasodilation 58, 97, 133
neuroimaging technique 162
neuroinflammatory disorder 119
neurokinin A 3, 93, 161
neurokinin B 93
neurokinin-1 receptor 140
neurokinin-1-receptor antagonist 57, 78, 124
neurokinin-1-receptor knockout mice 124
neuronal dysfunction 119
neuropeptide, sensory 119, 124
neuropeptide γ 93
neuropeptide K 93
neuropeptide Y-1 75
neurosensitizing molecule 119
NF-κB 149
N^{G}-nitro-L-arginine 122
nitric oxide 59, 70, 101, 119, 137
nitroglycerin 69, 101
non-prescription medication 25, 27
non-steroidal anti-inflammatory analgesic 71, 149
nucleus caudalis 124

occipital cortex 8
oligemia 8
otic ganglion 75

pain medication 25
pain threshold 3–5, 18
pain transmission 124
parasympathetic nervous system 72, 73
parasympathetic sphenopalatine ganglion stimulation 121
peptidergic nerve fiber 115
periaqueductal gray 14
peri-infarct depolarization 80
peripheral nerve fiber 124
phonophobia 12
photophobia 12
physical stress 119
pizotifen 150
plasma-protein extravasation 67, 75, 92, 124, 133, 137, 140
PNU-142633 134–136
positron emission tomography 12–14
postjunctional inhibitor 133
prejuctional inhibitor 133, 134
pro-inflammatory action 122
propranolol 147
prostaglandin 72, 100
prostaglandin E2 72
prostanoid receptor EP3 101

quality of life 24, 25

race 22
reactive hyperemia 8
reactive vasodilation 10, 11
relaxation therapy 27
retina 75, 162
rizatriptan 17, 123
RO-470203 137, 138

SB-220453 137
scanning electron microscopy 115
scintillating scotoma 6, 7
selective serotonin re-uptake inhibitor 151
selective serotonin-1D-receptor agonist 134
selective serotonin-1F-receptor agonist 135
self-relaxation 123
sensitization 77, 161
sensory neuron 119
sensory neuropeptide 119, 124
sensory nucleus of the trigeminal nerve 124
serotonin 73, 96, 124
serotonin-1A receptor 98, 146
serotonin-1B receptor 58, 70, 73, 97, 134
serotonin 1B/D-receptor agonist 58, 134
serotonin-1D receptor 134
serotonin-1F receptor 70, 97
serotonin-1F-receptor agonist 97
serotonin-2B receptor 73, 98, 146
serotonin 1-like receptor 17
serotonin receptor 17, 58, 70, 73, 97, 98, 134, 146
serotonin-7 receptor 146
single-photon emission computed tomography 76
social causation 23
social selection 24
socioeconomic status 23, 24
somatostatin 101
specialist consultation 26
sphenopalatine ganglion 75
spreading depression 6, 10, 12, 119, 124, 137
spreading excitation 10, 12
stress 119, 124
stress hormone 124
stroke 23
substance P 2, 55, 68, 93, 96, 115, 133, 140, 161
sumatriptan 12, 17, 18, 97, 123, 133
sumatriptan analogue 135
superficial temporal artery 2, 3, 14, 16, 17,
superior sagittal sinus 71, 77
sympathetic ganglion 122
sympathetic nervous system 2, 12

tachykinin 93
99-technetium labeled albumin 76
99-technetium-gluceptate 120

temporal artery 2
tension-type headache 23
T lymphocyte 122
tottering mouse 80
transformed migraine 22
tricyclic antidepressant 151
trigeminal ganglion 58, 119, 121
trigeminal ganglion stimulation 121
trigeminal nerve, sensory nucleus 124
trigeminal nerve fiber 65, 161
trigeminal system 91
trigeminovascular system 91
triptan 17, 18, 70
triptan, modes of action 17
tumor necrosis factor (TNF) 119

umbilical vein 122
unemployment 23
urine 123
urocortin 120

valproate 151
vanilloid-1 receptor 101
vascular permeability 119
vasoactive intestinal polypeptide 72, 75, 119
vasodilation 1, 2, 6, 7, 10–12, 58, 97, 119, 133
VCAM-1 152
verapamil 152
visual cortex 9, 11
visual migraine aura 80

4991W93 59, 134, 135

Xenon-clearance technique 7, 8

zolmitriptan 17

The PIR-Series
Progress in Inflammation Research

Homepage: http://www.birkhauser.ch

Up-to-date information on the latest developments in the pathology, mechanisms and therapy of inflammatory disease are provided in this monograph series. Areas covered include vascular responses, skin inflammation, pain, neuroinflammation, arthritis cartilage and bone, airways inflammation and asthma, allergy, cytokines and inflammatory mediators, cell signalling, and recent advances in drug therapy. Each volume is edited by acknowledged experts providing succinct overviews on specific topics intended to inform and explain. The series is of interest to academic and industrial biomedical researchers, drug development personnel and rheumatologists, allergists, pathologists, dermatologists and other clinicians requiring regular scientific updates.

Available volumes:

T Cells in Arthritis, P. Miossec, W. van den Berg, G. Firestein (Editors), 1998
Chemokines and Skin, E. Kownatzki, J. Norgauer (Editors), 1998
Medicinal Fatty Acids, J. Kremer (Editor), 1998
Inducible Enzymes in the Inflammatory Response, D.A. Willoughby, A. Tomlinson (Editors), 1999
Cytokines in Severe Sepsis and Septic Shock, H. Redl, G. Schlag (Editors), 1999
Fatty Acids and Inflammatory Skin Diseases, J.-M. Schröder (Editor), 1999
Immunomodulatory Agents from Plants, H. Wagner (Editor), 1999
Cytokines and Pain, L. Watkins, S. Maier (Editors), 1999
In Vivo *Models of Inflammation*, D. Morgan, L. Marshall (Editors), 1999
Pain and Neurogenic Inflammation, S.D. Brain, P. Moore (Editors), 1999
Anti-Inflammatory Drugs in Asthma, A.P. Sampson, M.K. Church (Editors), 1999
Novel Inhibitors of Leukotrienes, G. Folco, B. Samuelsson, R.C. Murphy (Editors), 1999
Vascular Adhesion Molecules and Inflammation, J.D. Pearson (Editor), 1999
Metalloproteinases as Targets for Anti-Inflammatory Drugs, K.M.K. Bottomley, D. Bradshaw, J.S. Nixon (Editors), 1999
Free Radicals and Inflammation, P.G. Winyard, D.R. Blake, C.H. Evans (Editors), 1999
Gene Therapy in Inflammatory Diseases, C.H. Evans, P. Robbins (Editors), 2000
New Cytokines as Potential Drugs, S. K. Narula, R. Coffmann (Editors), 2000
High Throughput Screening for Novel Anti-inflammatories, M. Kahn (Editor), 2000
Immunology and Drug Therapy of Atopic Skin Diseases, C.A.F. Bruijnzeel-Komen, E.F. Knol (Editors), 2000
Novel Cytokine Inhibitors, G.A. Higgs, B. Henderson (Editors), 2000
Inflammatory Processes. Molecular Mechanisms and Therapeutic Opportunities, L.G. Letts, D.W. Morgan (Editors), 2000

Cellular Mechanisms in Airways Inflammation, C. Page, K. Banner, D. Spina (Editors), 2000
Inflammatory and Infectious Basis of Atherosclerosis, J.L. Mehta (Editor), 2001
Muscarinic Receptors in Airways Diseases, J. Zaagsma, H. Meurs, A.F. Roffel (Editors), 2001
TGF-β and Related Cytokines in Inflammation, S.N. Breit, S. Wahl (Editors), 2001
Nitric Oxide and Inflammation, D. Salvemini, T.R. Billiar, Y. Vodovotz (Editors), 2001
Neuroinflammatory Mechanisms in Alzheimer's Disease. Basic and Clinical Research, J. Rogers (Editor), 2001
Disease-modifying Therapy in Vasculitides, C.G.M. Kallenberg, J.W. Cohen Tervaert (Editors), 2001
Inflammation and Stroke, G.Z. Feuerstein (Editor), 2001
NMDA Antagonists as Potential Analgesic Drugs, D.J.S. Sirinathsinghji, R.G. Hill (Editors), 2002

MIX
Papier aus verantwortungsvollen Quellen
Paper from responsible sources
FSC® C105338

If you have any concerns about our products,
you can contact us on
ProductSafety@springernature.com

In case Publisher is established outside the EU,
the EU authorized representative is:
Springer Nature Customer Service Center GmbH
Europaplatz 3, 69115 Heidelberg, Germany

Printed by Libri Plureos GmbH
in Hamburg, Germany